Gamma-ray Burst Correlations

Current status and open questions

Gamma-ray Burst Correlations

Current status and open questions

Maria Dainotti
Stanford University, California, USA
Jagiellonian University, Krakow, Poland
INAF, Bologna, Italy

IOP Publishing, Bristol, UK

ISBN   978-0-7503-1575-3 (ebook)
ISBN   978-0-7503-1573-9 (print)
ISBN   978-0-7503-1574-6 (mobi)

DOI   10.1088/2053-2563/aae15c

Version: 20190801

IOP Expanding Physics
ISSN 2053-2563 (online)
ISSN 2054-7315 (print)

British Library Cataloguing-in-Publication Data: A catalogue record for this book is available from the British Library.

Published by IOP Publishing, wholly owned by The Institute of Physics, London

IOP Publishing, Temple Circus, Temple Way, Bristol, BS1 6HG, UK

US Office: IOP Publishing, Inc., 190 North Independence Mall West, Suite 601, Philadelphia, PA 19106, USA

I dedicate the book to all people, family, friends and colleagues who have supported me in this long journey.

Contents

Preface

The intent of the book is to summarize the status of Gamma-ray Burst (GRB) correlations, the selection biases they undergo, their theoretical interpretation and their possible use as cosmological standard candles. The book is provided with a brief introduction on GRBs and some of the most popular theoretical modelling. Thus, the book is a useful compendium both for experts and for students who have just embarked in the study of correlations.

Acknowledgements

I thank R del Vecchio for fruitful discussions on how to organize the text material and for her initial participation in the book. The research leading to these results has received funding from the People Programme (Marie Curie Actions) of the European Union's Seventh Framework Programme FP7/2007-2013/ under REA grant agreement number 626267.

Author biography

Maria Dainotti

Maria Dainotti received her PhD in Relativistic Astrophysics from the University of Rome La Sapienza. She has since then focussed her research on high energy astrophysics, gamma-ray bursts and related astrophysical phenomena. Her research has led her to discover several important GRB relations bringing her name (the so-called Dainotti relations) and how to use GRBs as valuable cosmological tools. She is currently working as an American Astronomical Society International Chretienne fellow at Stanford University and concurrently serving as assistant professor at Jagiellonian University.

IOP Publishing

Gamma-ray Burst Correlations

Current status and open questions

Maria Dainotti

Chapter 1

Introduction

Gamma-ray bursts (GRBs) release in a few seconds the same amount of energy that the Sun releases in its entire lifetime, namely their isotropic emission is about 10^{48}–10^{55} erg (see Nakar 2007, Zhang 2011, Gehrels and Razzaque 2013, Berger 2014, Kumar and Zhang 2015, Mészáros and Rees 2015). GRBs were discovered by chance by the *Vela* satellites in the late 1960s and scientists immediately realized that GRBs were born outside our galaxy (Klebesadel *et al* 1973). Their isotropic angular distribution in the sky as well as an intensity distribution deviating from the $-3/2$ power-law (Paczynski 1991b, 1991a, Meegan *et al* 1992, Fishman and Meegan 1995, Briggs *et al* 1996) supported this conclusion. The first GRB with measured redshift validated these results. Indeed, this GRB was found at least at 2.9 Gpc (Metzger *et al* 1997) with $0.835 < z \lesssim 2.3$. Mazets *et al* (1981) found a bimodal distribution in the duration of GRBs detected by the Burst and Transient Source Experiment (BATSE) instrument (Meegan *et al* 1992). Since this discovery, GRBs have been divided into two classes: short (SGRBs) and long GRBs (LGRBs) with durations of $T_{90} < 2$ s and $T_{90} > 2$ s, respectively (Kouveliotou *et al* 1993). Moreover, Horváth (1998) and Mukherjee *et al* (1998) claimed the existence of a third class with intermediate durations. Later, Norris and Bonnell (2006) found a class of SGRBs with extended emission (SEE-GRBs) showing features common between SGRBs and LGRBs, but counter-arguments were recently discussed (Zitouni *et al* 2015, Tarnopolski 2017). Regarding the progenitor of LGRBs and SGRBs, two main scenarios are currently plausible: the collapse of a massive star connected to supernovae (SNe) for the first (Hjorth *et al* 2003, Malesani *et al* 2004, Woosley and Bloom 2006, Sparre *et al* 2011, Schulze *et al* 2014) and the coalescence of binary systems of neutron star–black hole (NS–BH) or NS–NS mergers (Eichler *et al* 1989, Paczynski 1991a, Narayan *et al* 1992) for the latter. In the latter case no connection to SNe has been claimed (Zhang *et al* 2009).

The fireball model (Cavallo and Rees 1978, Wijers *et al* 1997, Mészáros 1998, 2006) is one of the most famous models applied in the description of GRB emission.

In this model an ultra-relativistic and beamed electron/positron/baryon plasma is radiated by the compact central engine of either an LGRB or SGRB. The prompt emission phase, which consists of gamma-rays and hard X-rays, is generated from the collisions of blobs within the jet. In contrast, the afterglow phase, namely long broadband radiation (X-ray, optical, and radio) following the prompt phase, is created by the impact of the jet with the density medium. In the context of the fireball model, different x/gamma radiation processes are possible, such as synchrotron, inverse Compton (IC) and blackbody radiation, and sometimes a combination of these. Two types of fireball exist: the kinetic energy and the Poynting flux dominated (magnetic field dominated) fireball. Recently, Bégué and Pe'er (2015) investigated the Poynting flux dominated outflow that undergoes photospheric emission. The processes that characterize plasma thermalization lose their efficiency at a radius which has been indicated to be smaller that the photosphere radius by approximately two orders of magnitude. A combination of the Compton scattering below the photosphere and the conservation of the total number of photons imposes kinetic equilibrium between photons and electrons. Consequently, there is an increase of the photon temperature, which goes up to 8 MeV, specifically when the decoupling process of the plasma occurs in the photosphere. However, the model parameters do not strongly influence the results, even though they are not fixed parameters. Bégué and Pe'er (2015) indicated that the expected thermal luminosity, which is a tiny fraction of the total luminosity, could be observed. The magnetization of the outflows is constrained because the predicted peak energy is greater than the observed energy that characterizes the majority of GRBs. Continuing on this research theme, Pe'er (2015) summarizes new ideas, methods and instruments which have revolutionized the way in which prompt emission is studied. In particular, the author explains the latest observational outcomes and up-to-date theoretical interpretation of the same. Time-resolved spectral analysis has been demonstrated as a huge step forward from the observational perspective. This analysis has resulted in the discovery of a distinct high-energy component that leads to the delayed observation of the GeV photons, and strong evidence regarding the occurrence of the thermal component in many bursts. These outcomes led to numerous theoretical efforts that were meant to highlight the physical conditions characterizing the internal regions of the jet that emits the prompt photons and is responsible for the observed different spectral features. The main results of Pe'er's (2015) review are: (a) the clarification of the magnetic fields which determine the GRB spectra and their outflow; (b) the comprehension of the process of particle acceleration to higher energies in magnetic reconnection layers as well as shock waves; (c) the study of the dissipation of sub-photospheric energy broadening the Planck spectrum; and (d) the geometric effects deriving from aberration of the light.

Difficulties in matching the fireball model with observations are found because it is not possible to distinguish among the radiation processes or between the two kinds of fireball (kinetic energy or pointing flux dominated). An additional problem is that a more precise estimate of the jet opening angles and of the jet structure is needed. Additionally, Willingale *et al* (2007), hereafter W07, pointed out some issues in describing the light curves within this model. Indeed, the observed afterglow phase appears to be in agreement with the model for only ∼50% of GRBs. The discrepancy with the standard fireball model emerged when the *Swift* satellite was launched in 2004. For a brief summary of the current missions, see section 1.3. Rapid observation of the afterglow phases in several wavelengths was provided by *Swift*, with better coverage than earlier missions. *Swift* unveiled a more complicated light curve trend (O'Brien *et al* 2006, Sakamoto *et al* 2007, Zhang *et al* 2007) compared to previous studies. Nousek *et al* (2006) showed that GRBs often follow 'canonical' light curves which consists of several segments. The second segment, in the case that it is flat, is called plateau phase and can be described as being generated from accretion onto a BH (Cannizzo and Gehrels 2009, Cannizzo *et al* 2011, Kumar *et al* 2008), the evolution of a top-heavy jet (Duffell and MacFadyen 2015), or the delayed injection of rotational energy ($\dot{E}_{rot} \sim 10^{50}$–$10^{51}$ erg s^{-1}) from a fast rotating magnetar (Usov 1992, Zhang and Mészáros 2001, Dall'Osso *et al* 2011, Metzger 2011, Rowlinson and O'Brien 2012, Rowlinson *et al* 2014, Rea *et al* 2015). Indeed, a new magnetar can be produced through the explosion of a massive star or from the coalescence of two NSs. These models are described in chapter 2.

In this framework the analysis of correlations amongst several physical quantities of the prompt and plateau phases is relevant to discriminate among the models presented in the literature. Corrections for selection biases of the phenomenological correlations can broaden our knowledge of the mechanism responsible for such emissions. The first observation of gravitational waves (GWs), GW150914, was carried out by the Laser Interferometer Gravitational Wave Observatory (LIGO; Abbott *et al* 2016). It was allegedly produced by the merger of two BHs with masses (in units of solar masses) of $36^{+5}_{-4}M_\odot$ and $29^{+4}_{-4}M_\odot$, and represents an extremely significant discovery. It became even more important due to the analysis by Connaughton *et al* (2016) who announced a weak transient emission lasting 1 s and observed by *Fermi*/Gamma-Ray Burst Monitor (GBM) (Narayana Bhat *et al* 2016) almost in coincidence with GW150914; this emission was named GW150914-GBM. The false alarm probability of this event is 0.0022. Although the directions of these GW and GRB events are compatible, their association is still tentative due to the large errors in their estimated positions. In addition, SGRBs seem to be produced by NS–NS or NS–BH mergers, thus this connection is surprising. In addition, no signals of a compatible GRB were discovered by either *INTEGRAL* (Savchenko *et al* 2016) or by *Swift* (Evans *et al* 2016). Nevertheless, this discovery has already produced several scenarios to describe how a BH–BH merger can become a GRB: Li *et al* (2016) proposed that in a dense medium (see also Loeb 2016) GRBs are created by a BH via accretion of a mass $\simeq 10^{-5}M_\odot$ or, as described by Perna *et al* (2016), an SN explosion of two high-mass, low-metallicity stars and the matter emitted can produce an accretion disk and following on an SGRB. The observation of an afterglow phase visible months after the GW (Morsony *et al* 2016) could help in better understanding the connection between GWs and SGRBs. Another GW for which the association with a short GRB is evident and has been explained with the merger of a double NS is the case of GW 170817. For this event an X-ray counterpart has been discovered (Troja *et al* 2017). The X-ray and radio observations suggested an SGRB viewed off-axis.

Table 1.1 (mostly taken from Dainotti *et al* (2018) and Dainotti and Del Vecchio (2017)) presents the abbreviations and acronyms adopted throughout the book.

Table 1.1. Table with abbreviations from Dainotti and Del Vecchio (2017).

Abbreviation	Meaning
a	Normalization of the correlation
b	Slope of the correlation
b_{int}	Intrinsic slope of the correlation
BH	Black hole
CL	Confidence level
DE	Dark energy
EoS	Equation of state
E4	Sample with $\sigma_E = (\sigma^2_{\log L_{X,a}} + \sigma^2_{\log T^*_{X,a}})^{1/2} < 4$
E0095	Sample with $\sigma_E = (\sigma^2_{\log L_{X,a}} + \sigma^2_{\log T^*_{X,a}})^{1/2} < 0.095$
FS	Forward shock
Γ	Lorentz factor
GFR	GRB formation rate, the number of GRBs as a function of the redshift
H_0	Present-day Hubble constant
h	Hubble constant divided by 100
HD	Hubble diagram
IC	Intermediate class GRB
LGRB	Long GRBs
NS	Neutron star
Ω_M	Matter density in ΛCDM model
Ω_Λ	Dark energy density in ΛCDM model
Ω_k	Curvature in ΛCDM model
RS	Reverse shock
SN(e)	Supernova(e)
SGRB	Short GRB
SFR	Star formation rate
$\sigma_{\log L_{X,a}}$	Error in the luminosity
$\sigma_{\log T^*_{X,a}}$	Error in the time
σ_{int}	Intrinsic scatter of the correlation
ULGRB	Ultra-long GRB
V	Variability of the GRB light curve
W07	Willingale *et al* (2007)
w_0, w_a	Coefficients of the DE EoS $w(z) = w_0 + w_a z(1 + z)^{-1}$
XRFs	X-ray flashes
z	Redshift

1.1 The phenomenology of GRBs

Despite the fact that a very large variety of GRBs with different features has now been observed, the theoretical efforts on the existing models in the literature are devoted to identifying the common properties of a canonical GRB.

The first GRB peculiarity is their enormous observed fluence, between 10^{-7} and 10^{-5} erg cm^{-2}. Usually, the prompt high-energy emission, the main event, is observed in the gamma-rays, lasting from fractions of seconds to tens of seconds, followed by its counterpart, a long lasting multi-wavelength emission called the afterglow phase. The gamma-ray duration ranges from 10^{-3} to 10^3 s.

The GRB light curve components are visually introduced to acquaint the reader. In figure 1.1, the light curve of GRB 080430 observed by *Swift* is shown as a representative example displaying the prompt and afterglow phases.

Another important feature of GRB light curves is the variability, V, defined in section 3.2. Recent results on variability are discussed in Golkhou and Butler (2014) and Golkhou *et al* (2015). In the first paper, the authors constrained the lowest variability timescales of 938 GRBs observed up to July 2012, using the *Fermi*/GBM. It has been shown that the variability timescale in the soft band (or hard X-rays observed by *Swift*) is two to three times longer than for the hardest bands. Taking into account the upper limits and the detections, it has been discovered that less than 10% of the GRBs presented variability less than 2 ms. These particular timescales

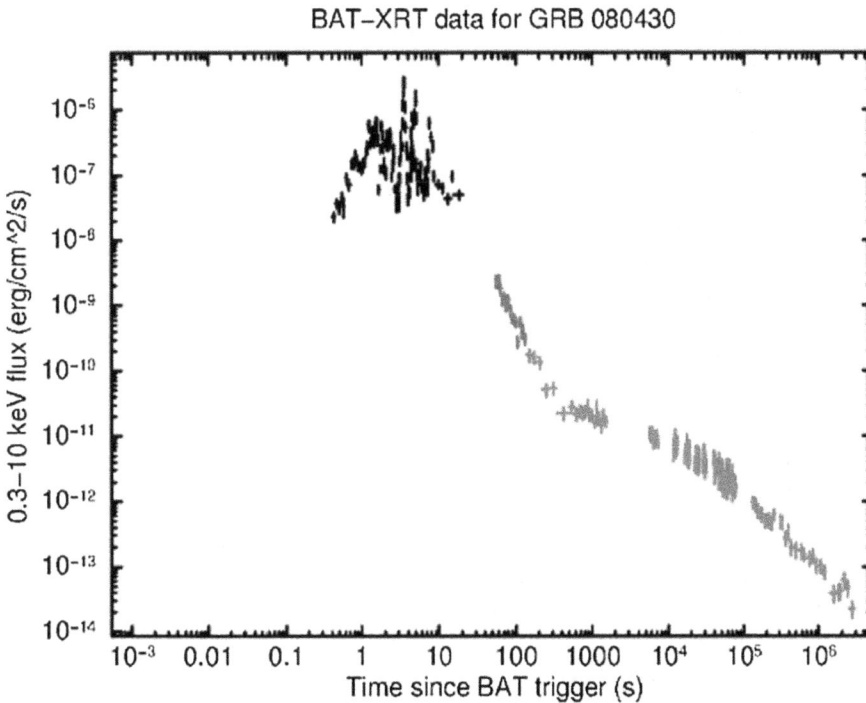

Figure 1.1. An example of a GRB light curve: GRB 080430 (from http://www.swift.ac.uk/burst_analyser). Credit: The Neil Gehrels Swift Observatory.

need the Lorentz factor of the emitting source, Γ, to be $\geqslant 400$, as well as typical emission radii with $R \equiv 10^{14}$ cm for LGRBs and $R \equiv 3 \times 10^{13}$ cm for SGRBs. Golkhou *et al* (2015) established the minimum variability timescale that characterizes the light curves of GRBs using Haar wavelets. The employed approach takes the average of GRB data, determining a cumulative measure regarding signal variation whilst maintaining the sensitivity for narrow pulses present in complex time series. The technique has been applied to studying a huge sample of the GRB light curves observed by *Swift*. Given the large number of GRBs with known redshift, a new quantity has been defined called the minimum rest-frame variability timescale. Researchers discovered a minimum timescale for LGRBs of 0.5 s, whilst the shortest timescale discovered, implying a compact central engine, is 10 ms. Further implications for the GRB fireball model are presented together with a correlation between the redshift and minimum timescale, that may be partly caused by redshift evolution.

1.1.1 The prompt emission

The prompt emission is the phase in which the gamma-ray instrument detects a signal above the background in a statistically meaningful way (Piran 2004). Usually, lower-energy emission (X-ray, optical) occurs simultaneously with gamma-ray emission. Sometimes such an X-ray signal ends up being more intense than the gamma-ray one; bursts in which this situation occurs are called X-ray flashes (XRFs; Heise *et al* 2001), as explained in section 1.1.5.

The main feature of the prompt emission light curve is the great variety of temporal profiles, shown in figure 1.2. Indeed, it is possible to find very simple light curves with a fast rise exponential decay (FRED) structure or very complex multi-peaked structures on timescales up to $\delta t_a \approx 10$ ms, where δt_a is an observed single pulse width (Piran 2004). If one looks at the most complex light curves, they seem to be composed of individual pulses, which are the fundamental elements of the light curve (Piran 2004). Each pulse is characterized by: (a) a FRED structure (Norris *et al* 1996), also always present in simple profiles, (b) a spectral trend where the peak energy declines as a function of the photon fluence in an exponential way (Norris *et al* 1996) and (c) low-energy emission delayed compared to high-energy emission (Norris *et al* 1996).

1.1.2 The afterglow emission

After the occurrence of prompt emission, the afterglow multi-wavelength emission begins. The overall energy emitted in this stage is generally a few percent of the GRB total energy. The X-ray afterglow phase usually starts from hundreds of seconds to several hours after the prompt emission. In this region the X-ray afterglow fluxes show a phenomenological relationship, both with the frequency ν and the observed time t (Piro *et al* 2000): $f_\nu(t) \propto \nu^{-\beta} t^{-\alpha}$, with $\alpha \sim 1.4$ and $\beta \sim 0.9$.

Together with this X-ray emission, the afterglow phase may also be observed in the optical and IR bands. Around one day after the burst, the brightness of the optical afterglow phase is about 19–20 mag. In the beginning, the light curve

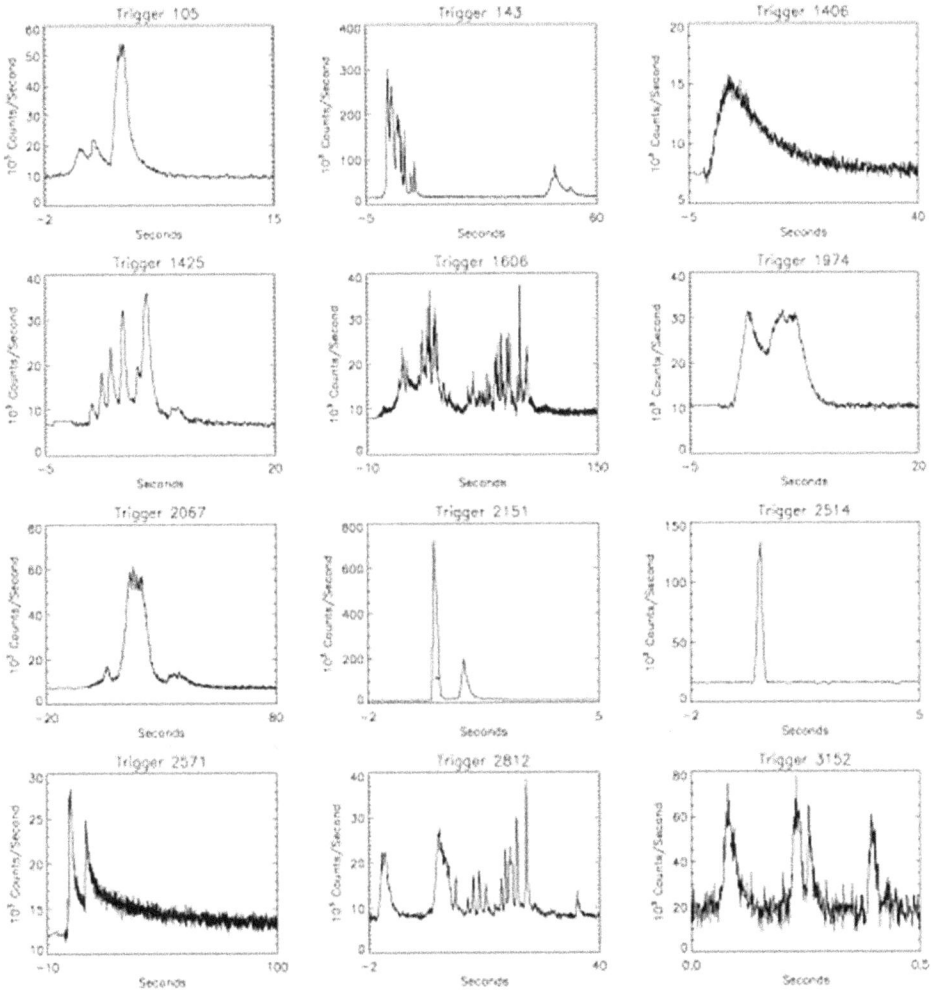

Figure 1.2. Samples of prompt emission light curves observed by *Swift*. Credit: The Neil Gehrels Swift Observatory.

declines as a power-law, $t^{-\alpha}$ with $\alpha \approx 1.2$, but with large variations around this number. The optical afterglow phases show behaviour similar to the X-ray phase only in some cases, as shown by Nardini *et al* (2006).

Many afterglow light curves at later times show a change in the power-law index to a steeper decline with $\alpha \approx 2$. Such a break in the afterglow phase is usually chromatic (dependent on the frequency), but in a few cases it has been observed in different energy bands. A classic example of such a situation was seen in GRB 990510. Usually, the break is fitted with a phenomenological formula: $F_\nu(t) = f_*(t/t_*)^{-\alpha_1}(1 - \exp[-(t/t_*)^{(\alpha_1-\alpha_2)}](t/t_*)^{(\alpha_1-\alpha_2)})$ (Piran 2004). This break is known as a jet break and the time at which such a break appears can be applied to calculate the jet opening angle (Rhoads 1997) or the observing angle of the standard jet model.

With the launch of *Swift* a new era started in the study of GRBs. *Swift*'s primary goal is not only to detect a large number of GRBs so that we can achieve a statistically meaningful sample, but also to retrieve data in the bands between 0.3–10 keV, and between 1700 and 6500 Å after the initial few tens of seconds after the trigger. The result is that the afterglow light curve shows a more complex behaviour than the simple power-law decay observed in the pre-*Swift* era. The up-to-date *Swift* light curves in the X-ray range can be divided in two morphological types (Chincarini *et al* 2007): one starting with a very steep light curve decay and the other showing a flat decline. The first class is preferred for the majority of GRBs, displaying a 'canonical' behaviour (Nousek *et al* 2006, O'Brien *et al* 2006, Sakamoto *et al* 2007): after the initial steep decay ($F \propto t^{-\alpha_1}$, with $3 \leqslant \alpha_1 \leqslant 5$) the light curve presents a flat decaying phase ($F \propto t^{-\alpha_2}$, with $0.5 \leqslant \alpha_2 \leqslant 1$) followed by a steeper decay ($F \propto t^{-\alpha_3}$, with $1 \leqslant \alpha_3 \leqslant 1.5$), consistent with those seen in previous missions and, in some cases, overlaid X-ray flares (see figure 1.3). The spectrum, except in a few cases, remains constant throughout all these stages of the afterglow phase in the sample used by Nousek *et al* (2006). However, spectral evolution of the afterglow was presented later (Evans *et al* 2009) when more data became available for a statistically significant sample. It is worth noting that the final break has been seen in only less than ~10% of the afterglow phases followed by *Swift*. Even if the last part of the GRB prompt emission seems to be represented by the steep α_1 segment, shown

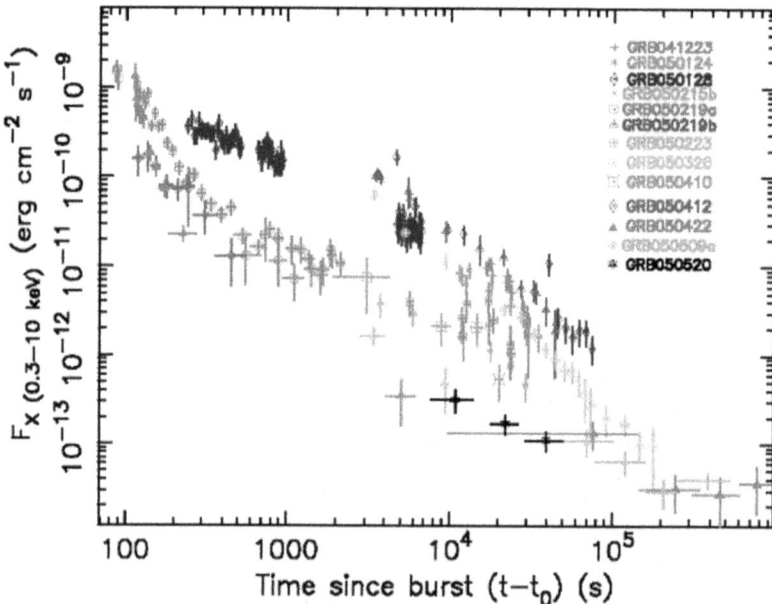

Figure 1.3. Example of GRB afterglow light curves observed by *Swift*, showing the two different trends reported by Chincarini *et al* (2007). Reproduced with permission from Nousek *et al* (2006). Copyright 2006 The American Astronomical Society.

in the canonical behaviour, Liang and Zhang (2005) and Sakamoto *et al* (2007) stated that the origin of the transition phase from a shallow α_2 to a steeper α_3 needs further analysis to be fully explained.

In the *Swift* XRT light curves flares are present (Nousek *et al* 2006, see figure 1.4). In some of them the flux varies on timescales $\delta t_a \ll T_{X,a}$ with a very steep rise and decay. In the case where the flare is luminous enough, its spectral index is not in agreement with the power-law behaviour exhibited by the X-ray light curve without the flares.

1.1.3 The broadband spectrum

An important physical characteristic of GRB phenomenology is the emission energy spectrum (see figure 1.5). Usually, LGRBs have spectral peak energy around 150 keV, a low-energy spectral slope around -1 and a high-energy slope around -2 (Band *et al* 1993). Instead, SGRBs often have a greater spectral energy peak and a low-energy spectral slope around -0.5. Originally, the GRB spectra had been fitted employing the Band function (Band *et al* 1993):

$$N_E(E) = A_{\mathrm{norm}} \times \begin{cases} \left(\dfrac{E}{100 \text{ keV}}\right)^{\alpha} \exp\left(-\dfrac{E}{E_0}\right), & E \leqslant (\alpha - \beta)E_0 \\[4mm] \left[\dfrac{(\alpha - \beta)E_0}{100 \text{ keV}}\right]^{\alpha - \beta} \left(\dfrac{E}{100 \text{ keV}}\right)^{\beta} \exp(\alpha - \beta), & E \geqslant (\alpha - \beta)E_0 \end{cases}, \quad (1.1)$$

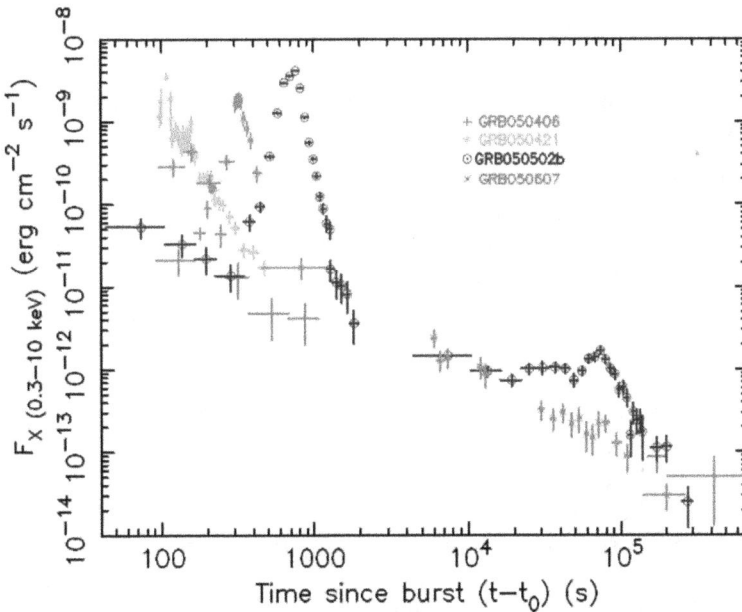

Figure 1.4. Examples of GRB afterglow light curves with X-ray flares. Reproduced with permission from Nousek *et al* (2006). Copyright 2006 The American Astronomical Society.

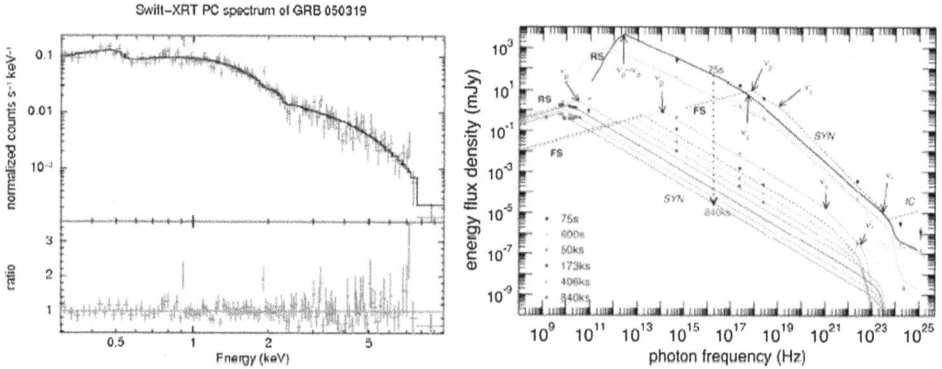

Figure 1.5. Left panel: an example of a spectrum of GRB 050319 taken from the *Swift* XRT repository. Credit: The Neil Gehrels Swift Observatory. Right panel: a model of broadband GRB spectra. Solid lines are for the reverse shock (RS) emission while dashed lines show the forward shock (FS). Data points are provided for different times: 75 s (black), 600 s (pink), 50 000 s (red), 173 000 s (blue), 406 000 s (green) and 840 000 s (purple). Reproduced from Panaitescu *et al* (2013). By permission of Oxford University Press on behalf of the Royal Astronomical Society.

where A_{norm} is the normalization, α and β are the low- and high-energy exponents of the Band function, respectively, E_0 is the spectral break energy and $N_E(E)$ is the photon flux density computed in photons $cm^{-2}s^{-1}keV^{-1}$. Currently, additional components need to be considered in spectral modelling: a high-energy power-law or exponentially attenuated power-law component (González *et al* 2003), and a blackbody or photospheric component (Ryde *et al* 2006) to describe the thermal radiation from the photosphere.

1.1.4 The observational classification: long, short, intermediate class and ultra-long GRBs

This section describes in more detail the problem of GRB classification mentioned in section 1. LGRBs and SGRBs appear to have very different observational properties beyond the difference in duration (see figure 1.6). It was concluded that SGRBs have a harder spectrum than LGRBs, and this enforced the idea that SGRBs were a different entity, instead of being, for example, single peaks of LGRBs hidden by noise. From the analysis of unimodal duration distributions (Horváth 1998, 2002, Horváth *et al* 2008, Horváth 2009, Huja *et al* 2009, Řípa *et al* 2009) and higher dimensional parameter spaces (Mukherjee *et al* 1998, Horváth *et al* 2006, Řípa *et al* 2009, Horváth *et al* 2010, Veres *et al* 2010, Koen and Bere 2012) an intermediate class of GRBs was introduced. This class was later investigated further (Norris and Bonnell 2006). The observed afterglow phases of this intermediate class are characterized by a soft long bump following the prompt emission, and by a temporal evolution at late time analogous to the LGRBs, showing a 'canonical' behaviour and, in several cases, the presence of X-ray flares (Barthelmy *et al* 2005). These bursts are harder than LGRBs (Amati 2006). Moreover, they also appear to be closer than what was expected before. In fact, the average redshift is $z \sim 0.4$,

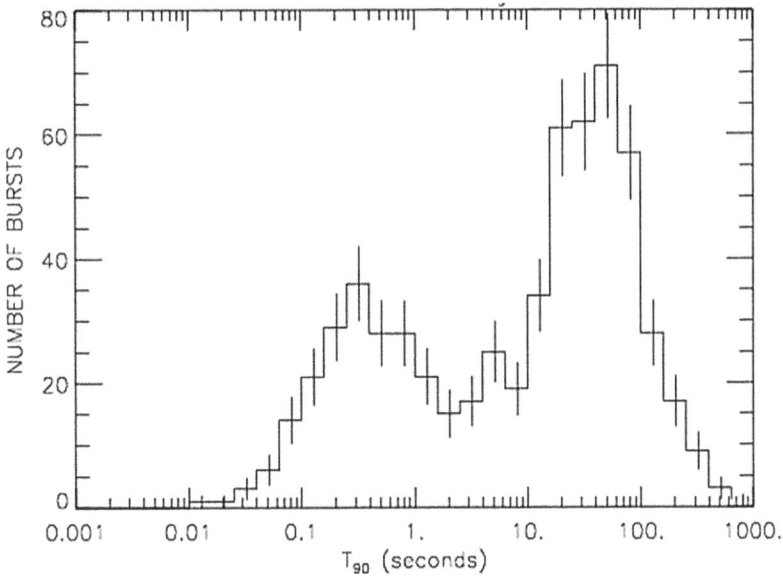

Figure 1.6. Histogram of the duration of 1234 GRBs observed by BATSE as shown in the BATSE 4B Catalog. Reproduced with permission from Meegan *et al* (1998). Copyright 1998 The American Institute of Physics.

much smaller than the average z of the LGRBs detected by *Swift*, $z \sim 2.3$ (Guetta 2006) and, therefore, their luminosity is lower and their local rate is higher (Guetta 2006, Barthelmy *et al* 2005, Fox *et al* 2005). All these recent observations support the idea that these bursts belong to a different population than the LGRBs (Fox *et al* 2005). In particular, it has been supported by Barthelmy *et al* (2005) that their association with non-star-forming host galaxies indicates that the collapse of a massive star cannot generate this intermediate class of GRBs. On the other hand, the similarity of all their properties definitely indicates commonalities in the origin of SGRBs (Fox *et al* 2005). Indeed, their origin seems to be consistent with the merging of a compact object binary (Fox *et al* 2005), even if the NS–NS binary merger models predict energy injection times much shorter than, for example, the 200 s observed for GRB 050724 (Barthelmy *et al* 2005). BH–NS mergers are more promising, but even these models cannot extend emission beyond a few tens of seconds (Barthelmy *et al* 2005). However, with the advent of the GW associated with SGRBs, the NS–NS scenario is confirmed.

Virgili *et al* (2013) and Levan *et al* (2014) claimed that GRBs with $T_{90} > 10000$ s (ULGRBs) are statistically separated from the classic LGRBs, thus they could constitute an additional class.

In previous years, distant star-forming galaxies seemed to produce LGRBs. Later on, associations with core collapse SNe were pointed out, but not for each LGRB (Fynbo *et al* 2006, Della Valle *et al* 2006). This indicates the existence of different types of progenitors for LGRBs. The metallicity Z is another characteristic worth considering when studying the progenitor system. Woosley and Bloom (2006) concluded that, in the collapsar model, LGRBs come from massive stars with $Z/Z_\odot$ lower than $\simeq 0.1$–0.3. Metal-rich systems are a suitable environment for many GRBs (Perley *et al* 2016). Greiner *et al* (2015) found that models other than the collapsar one are able to produce LGRBs and they are relevant to GRB studies.

1.1.5 The hardness ratio: GRBs, X-ray rich and X-ray flashes

The hardness ratio is given by the ratio of the flux of the third BATSE channel to the flux of the second BATSE channel (100–300 keV and 50–100 keV, respectively, see figure 1.7). Recently, a new phenomenon very similar to GRBs, called X-ray flashes (XRFs), was investigated (Barraud *et al* 2003). The only difference to GRBs is that XRF flux has its peak in the X-ray range, and it was clearly indicated that these two phenomena are connected. Strohmayer *et al* (1998) pointed out 22 such bursts detected by Ginga. Among the bursts detected by Ginga, the spectra of just 36 were found to be very soft, with a duration and spectral form similar to LGRBs, but with lower observed peak energy than BATSE bursts (Kaneko *et al* 2007, Sakamoto *et al* 2005). From wide-field camera (WFC) sources, a group of fast X-ray transients which last less than 10^3 s and were not observed by the GBM was found.

From the analysis of LGRBs and XRFs, it was noted that the spectral properties of X-ray afterglow phases for these two classes are compatible, but the temporal features are dissimilar. LGRB afterglow phases present a break at earlier times and with steeper indices than XRF ones; in addition, XRF afterglow luminosity is lower than that of LGRBs. To conclude, XRFs emit X-ray energy with a spatial

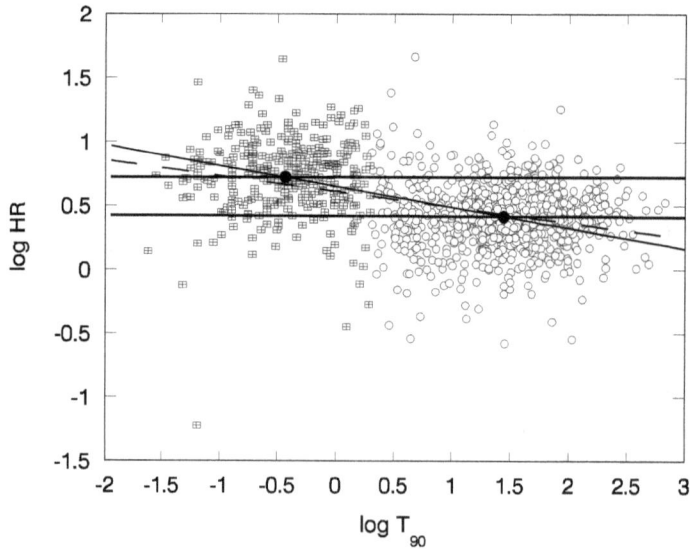

Figure 1.7. The hardness–duration correlation for BATSE GRBs. SGRBs (squares) seem to be harder than LGRBs (circles). The ratio of the flux of the third BATSE channel to the flux of the second BATSE channel (100–300 keV and 50–100 keV, respectively) is called hardness. Reproduced from Qin *et al* 2000. Copyright 2000 Oxford University Press. By permission of Oxford University Press on behalf of the Astronomical Society of Japan.

distribution, and spectral and temporal features comparable to LGRBs (Heise *et al* 2001, Kippen *et al* 2001). The feature discriminating XRFs from GRBs is the prompt emission spectrum, which for XRFs peaks at energies one order of magnitude smaller than the energies for LGRBs. In addition, XRFs have higher fluence between 2 and 30 keV in comparison to fluence in the gamma-ray band (30–400 keV). The current distinction is helpful for the study of GRB correlations, also because they can be modified by employing several GRBs classes (Amati 2006, Dainotti *et al* 2010). Between XRFs and GRBs there is an additional class called X-ray rich (XRR), characterized by very dim gamma to X-ray fluence in comparison to that of GRBs (D'Alessio and Piro 2005).

1.2 The phenomenological Willingale model

After the description of the properties of the prompt and afterglow phases, this section describes a phenomenological model introduced by W07 to characterize the plateau emission.

1.2.1 The functional form of X-ray decays

W07 studied several X-ray *Swift* light curves and claimed that they can be fitted with one or two parts described by an exponential decay plus a power-law decaying phase. The first part represents the prompt phase and the second part the afterglow phase. The functional form for both parts is given by

$$f_i(t) = \begin{cases} F_i e^{\alpha_i \left(1 - \frac{t}{T_i}\right)} e^{-\frac{t_i}{t}}, & t < T_i, \\ F_i \left(\frac{t}{T_i}\right)^{-\alpha_i} e^{-\frac{t_i}{t}}, & t \geqslant T_i, \end{cases} \tag{1.2}$$

where $i = p, a$, where p and a denote the prompt and afterglow phases, respectively. (T_i, F_i) is the point where the exponential phase changes into the power-law decaying phase; α_i defines the temporal decay index of the power-law, while t_i indicates when, in each component, the rise phase starts. The two-component model is shown in figure 1.8.

W07 fitted the light curves of 107 GRBs (see figure 1.9) assuming the form $f(t) = f_p(t) + f_a(t)$.

The prompt phase, mostly observed by BAT, takes place at $t < T_{X,p}$ (see section 3.2 for the definition), while the plateau phase, observed by XRT, occurs at $t < T_{X,a}$. In addition, the growing part of the afterglow phase is not visible because the prompt phase in the beginning always dominates. Even if bright flares are responsible for only ~10% of the total fluence, they were removed from the W07 fitting procedure.

From the above analysis, W07 derived the distribution of $T_{X,a}$ versus $T_{X,p}$ for those objects with two-component fits, pointing out no correlation between these two parameters. Studying the prompt and afterglow fluences, it was found that high prompt fluence implies high afterglow fluence, with large scatter, however.

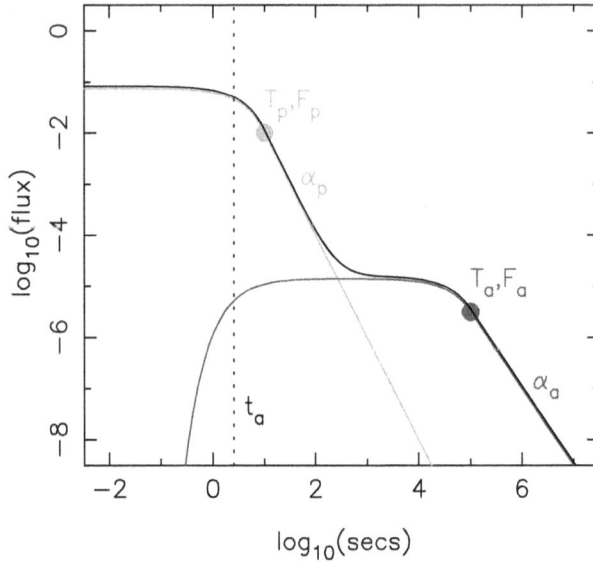

Figure 1.8. The functional form of the X-ray decay. The prompt component (green) has $t_p = 0$. The afterglow component (blue) begins at time t_a. log(flux) represents the quantity $f_i(t)$ from the equation (1.2) for the prompt and afterglow components, and log(secs) indicates the time. Reproduced with permission from W07. Copyright 2007 The American Astronomical Society.

In conclusion, applying the model by W07, the X-ray light curve can be reproduced well and many of the correlations between afterglow parameters are based on this phenomenological model.

1.3 The past and current missions observing GRBs

GRBs were detected by the *Vela* satellites, composed of X-ray (working in the 3–12 keV energy range) and gamma-ray (in the 150–750 keV range) detectors. Through the investigation of the different arrival times of GRB emission to each satellite, a terrestrial origin for these phenomena was ruled out. At the beginning of the *Compton Gamma Ray Observatory* mission (*CGRO*, Meegan *et al* 1992) in 1991, a systematic observation of GRBs could begin. The *CGRO* mission constituted four instruments (The Burst and Transient Source Experiment (BATSE), The Oriented Scintillation Spectrometer Experiment, The Compton Telescope and the Energetic Gamma Ray Experiment Telescope) and worked in the range between 30 keV and 30 GeV. Using BATSE, it was discovered that GRBs have no preferential direction in the sky (Meegan *et al* 1992, Efron and Petrosian 1995, Tegmark *et al* 1996)[1]. This space mission was running until the beginning of 2000 and it was one of the most decisive space missions at the time, providing the first survey of the whole sky above 100 MeV, an extensive survey of the Galactic Centre of the Milky Way and the discovery of the first soft gamma-ray repeaters (Bassani *et al* 1995). In particular, the

---

[1] For the most recent works on this topic, see Tarnopolski (2017) and Řípa and Shafieloo (2017).

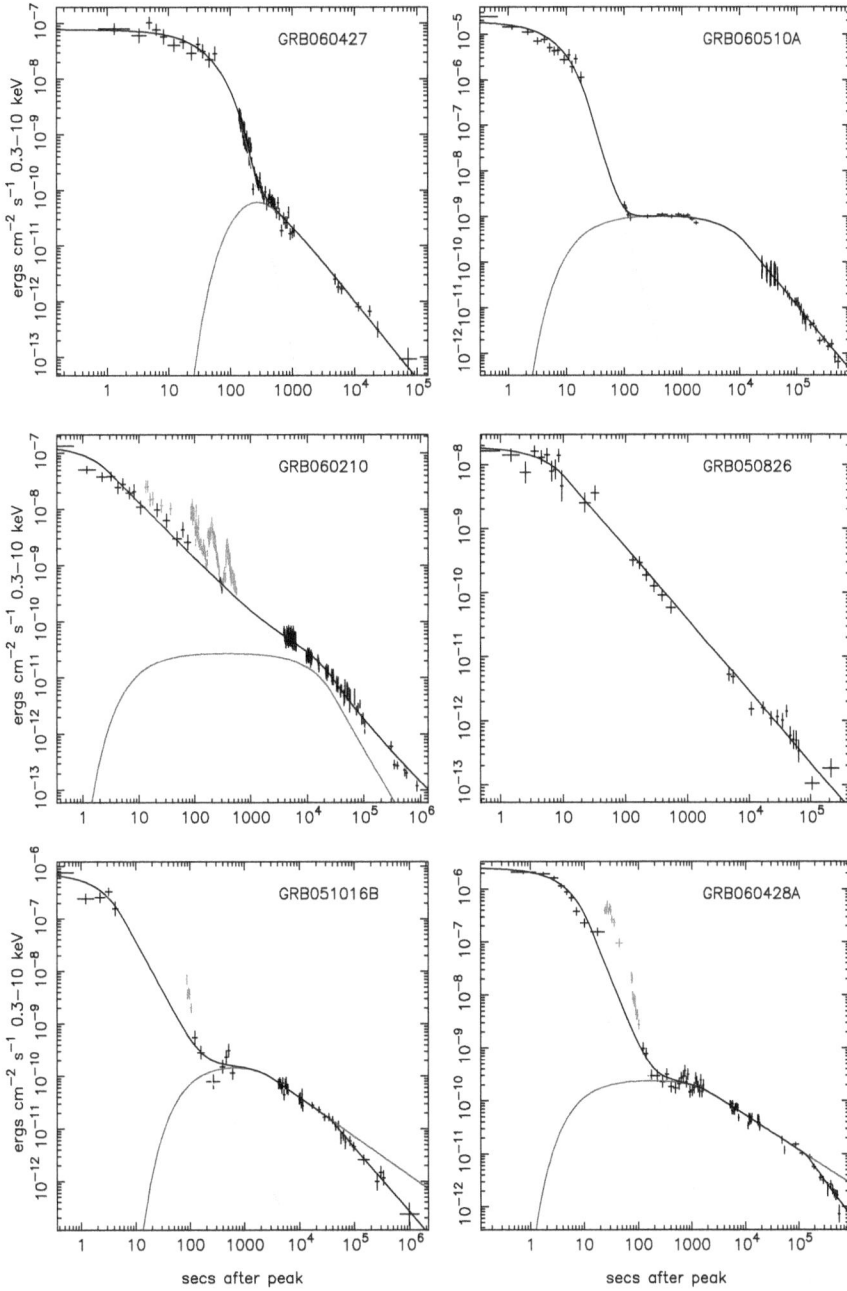

Figure 1.9. Several fits to X-ray decay curves. Both prompt and afterglow phases are represented similarly to figure 1.8. Top panels: an example of when the afterglow component is dominant at late times. Middle left panel: an example of when the afterglow phase displays a bump in the decline, but the prompt phase decay is dominant at late times. Middle right panel: a single power-law fit showing no afterglow phase. Bottom panels: two fits which include a late temporal break. Flares are shown in red and were excluded from the fitting. Reproduced with permission from W07. Copyright 2007 The American Astronomical Society.

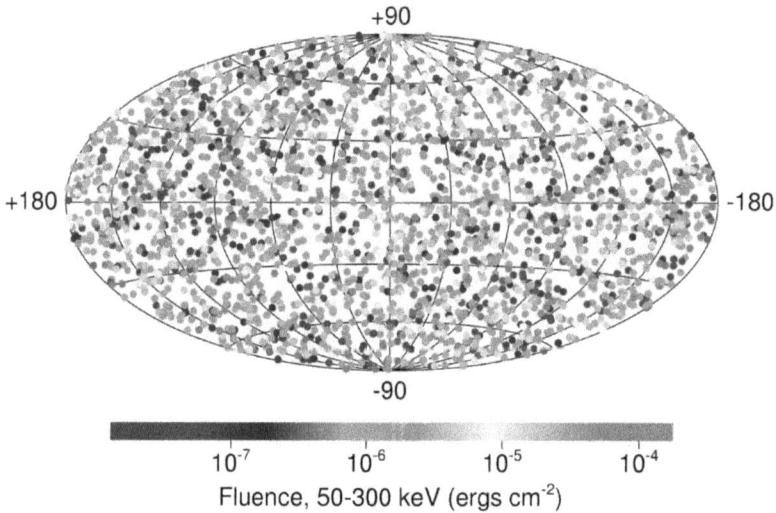

Figure 1.10. The distribution of the GRBs detected by BATSE (from http://www.swift.ac.uk/about/grb.php). The coloured bar denotes the GRB fluence scale. Credit: The Neil Gehrels Swift Observatory.

CGRO mission followed GRB 990123, one of the most luminous GRBs at the time, in great detail. From its analysis, it was confirmed that GRBs are highly collimated jet phenomena.

Figure 1.10 displays the fluence (energy flow per unit area) of the GRBs detected by BATSE. The purple points mark the lowest fluence values ($\sim 10^{-7}$ erg cm^{-2}), while the red dots denote the highest fluence values reaching scales at $\sim 10^{-4}$ erg cm^{-2}.

Later, in 1996, an Italian-Dutch space mission *BeppoSAX* (Boella *et al* 1997) was launched to explore GRBs. In six years of operation, the precision of its observations in the energy range between 0.1 and 300 keV confirmed the extra-galactic origin of GRBs (Costa *et al* 1997). *BeppoSAX* was composed of four X-ray telescopes with two spectrometers (a low-energy concentrator spectrometer and medium energy concentrator spectrometer), a high-pressure gas scintillator proportional counter, the photo-switch detection system and the wide-field camera. As shown in section 1.1.1, GRB light curves are composed of the prompt phase followed by a decaying afterglow phase. For the first time, in February 1997, the *BeppoSAX* satellite detected the afterglow phase as fading X-ray emission related to GRB 970228 (Costa *et al* 1997). It was followed, 20 h after the GRB, by the optical counterpart detected by the William Herschel Telescope[2] (Groot *et al* 1997).

In 1999 the *Chandra X-Ray Observatory*[3] was launched and it is still operating. It carries the following instruments: the AXAF charged coupled imaging spectrometer, the high-resolution camera, the high-energy transmission grating, and the low-energy transmission grating, all working in the 0.1 to 10 keV X-ray energy range.

---

[2] http://www.ing.iac.es/Astronomy/telescopes/wht/.
[3] http://chandra.harvard.edu/about/specs.html.

With its high resolution, large collecting area and sensitivity to the hard X-ray energy band, *Chandra* is useful for the analysis of GRB spectral emission lines (Sako *et al* 2005), and for the investigation of the nature of GRB host galaxies (Fruchter 2000) and their star formation regions (Watson *et al* 2004). In particular, with the observation of GRB 991216, spectral X-ray emission lines were connected for the first time with a GRB (Piro *et al* 2000). This finding is important because it allowed, the estimation of the nature of the GRB progenitor in more detail.

Also in 1999, the *XMM-Newton* space mission[4] was launched (Jansen *et al* 2001) and it is still obtaining important results on X-ray sources, carrying out narrow and broad range spectroscopy, and providing X-ray and optical (0.1–15 keV and 180–650 nm, respectively) imaging for each object. The main instruments of this mission are: the European Photon Imaging Camera (EPIC) Metal-Oxide–Silicon, the EPIC-PN, the reflection grating spectrometer and the optical monitor. In particular, it observed GRB 001025A to investigate the so-called dark bursts. These bursts are optically subluminous GRBs with an optical-to-X-ray spectral index, $\beta_{OX,a}$, less than 0.5 (Jakobsson *et al* 2004). Furthermore, *XMM-Newton* observed GRB 011211 and GRB 030227 to study the X-ray emission lines (Reeves *et al* 2002, Watson *et al* 2003), and GRB 030329 to investigate the late afterglow phase (Tiengo *et al* 2003).

In 2000, the *High Energy Transient Explorer 2* (*HETE-2*) was launched[5] (Shirasaki *et al* 2003) after the failed launch in 1996 of the first *High Energy Transient Explorer*. Until the end of the mission in 2008, its main goal was the multi-wavelength investigation of GRBs and the precise detection of their positions with ~10 arcsec accuracy. The mission included two X-ray detectors (a soft X-ray camera (SXC) and a wide-field X-ray monitor (WXM)) working in the range 0.5–14 keV and 2–25 keV, respectively, and four gamma-ray detectors (FREGATE) operating in the 6–400 keV energy band. In particular, this mission detected GRB 030329, a reliable proof for the GRB–SN connection, and GRB 050709, the first SGRB with an observed optical counterpart. Then, due to the observations of *HETE-2*, it was found out that some dark GRBs fade in the optical energy band very rapidly, while others are dimmer, but still detectable with large telescopes. In addition, *HETE-2* discovered the existence of XRFs (see section 1.1.5 for details) and achieved an arcminute-precise positioning of GRBs within tens of seconds of the beginning of a GRB.

In 2002, the *INTEGRAL* space mission was launched (Teegarden and Sturner 1999) and it is still operative. Its main result is a sky map in the soft X-ray range (15 keV–10 MeV). *INTEGRAL* is composed of an imager (IBIS), a spectrometer (SPI), together with X-ray (JEM-X, 3–35 keV) and optical (OMC, 500–850 nm) detectors to provide simultaneous observations in these energy bands. In particular, the *INTEGRAL* mission investigated GRB 031203, one of the few GRBs connected to SNe at that time, and GRB 041219A, one of the longest and brightest GRBs ever observed at the time (Gotz 2013).

---

[4] http://sci.esa.int/xmm-newton/.
[5] http://space.mit.edu/HETE/spacecraft.html.

A major important space mission for investigating GRBs is the Neil Gehrels Swift Observatory (hereafter Swift) satellite (Gehrels *et al* 2004), which has been operating since 2004. It is composed of three instruments: the Burst Alert Telescope (BAT, 15–150 keV), the X-Ray Telescope (XRT, 0.3–10 keV) and the Ultraviolet/Optical Telescope (UVOT, 170–650 nm). As a result, the *Swift* space mission detected the furthest GRBs (GRB 090423 at $z = 8.26$ and GRB 090429B at $z = 9.2$) and one of the brightest GRBs ever observed (GRB 130427A, $z = 0.34$ and one-second peak photon flux measured by BAT to be 331 ph cm^{-2} s^{-1}). The analysis of the GRB light curves detected from *Swift* pointed out the complex and often different behaviour of these phenomena (Nousek *et al* 2006).

In 2007, the *Astro-Rivelatore Gamma a Immagini Leggero* (*AGILE*) mission was launched (Longo *et al* 2007), with the main aim of studying the gamma-ray sources in the Universe at high energies. In particular, it is investigating active galactic nuclei (AGNs), GRBs, X-ray and gamma galactic objects, and diffuse galactic and extra-galactic gamma-ray emissions. *AGILE* includes a gamma-ray imaging detector (GRID) working from 30 MeV to 50 GeV, a SuperAGILE (SA) hard X-ray monitor operating in the energy range between 18–60 keV, a mini-calorimeter (MCAL) running in the interval 350 keV–100 MeV and an anti-coincidence system (AC), consisting of a plastic scintillator, to support the removal of unwanted background events.

Another ongoing mission used for gamma-ray studies of GRBs is the *Fermi* gamma-ray space observatory launched in 2008 (Carson 2007). Its aim is to observe the Universe in the low- and medium gamma-ray energy ranges and it is composed of two instruments: the Large Area Telescope (LAT) for obtaining an all-sky survey investigating the high-energy emission between 20 MeV and 300 GeV, and the GBM for analysing GRBs in the energy band between 8 keV and 40 MeV. As a result, *Fermi* detected the strongest GRB at the time, GRB 080916C (Abdo *et al* 2009). In parallel with the *Swift* mission, it detected the even more energetic GRB 130427A[6] (Liu *et al* 2013). GBM was able to localize the SGRB associated with GW170817.

One should also mention the *NuSTAR* X-ray mission launched in 2012 (Hailey *et al* 2010). It works in the energy range between 3 and 79 keV and focuses, in particular, on high-energy X-ray spectroscopy of astrophysical sources. It mostly investigates the compact galactic objects radiating in the X-ray band and non-thermal radiation in young SN remnants using the excellent technology of solid state cadmium zinc telluride pixel detectors. In fact, *NuSTAR* has observed a few GRBs with data in other energy ranges also available from other telescopes, such as GRB 130427A and GRB 130925A. The wide-range data samples available for these GRBs allowed for their broadband spectral analysis.

Finally, the Indian *Astrosat* satellite[7], launched in 2015 (Agrawal 2005), is one of the most recent space missions capable of multi-wavelength observations. It is composed of the ultraviolet imaging telescope (UVIT), the soft X-ray imaging telescope (SXT), the LAXPC instrument, the cadmium zinc telluride imager (CZTI),

---

[6] https://http://www.nasa.gov/topics/universe/features/shocking-burst.html.
[7] http://astrosat.iucaa.in/.

the scanning sky monitor (SSM) and the charged particle monitor (CPM). These instruments cover the emission range from the far ultraviolet (130–180 nm) to the hard X-ray (10–150 keV) energy bands. Even though CZTI is a spectroscopic device, it is employed as a wide angle GRB monitor and it allows for spectro-polarimetric analysis of GRBs in the 100–300 keV band. The study of the polarization of the brightest 11 GRBs observed by the CZTI during its first year exhibited the importance of this satellite for the investigation of GRBs.

All the telescopes mentioned above are space missions, but there are also some ground based telescopes that are important in the study of GRBs. The main ground based telescopes are: the radio telescope Very Large Array (VLA) (Napier *et al* 1983), the Gamma-Ray Burst Optical/Near-Infrared Detector (GROND) (Greiner *et al* 2008) and the optical/infrared Very Large Telescope (VLT) (Dekker *et al* 2000).

VLA became operational in 1980. It is composed of 27 independent antennae working in the frequency range 74 MHz–50 GHz with an angular resolution between 0.04 and 0.2 arcsec. Its main goal is to investigate galaxies, stars, quasars, pulsars and GRBs. However, radio observations of GRBs from VLA are mostly used to support optical and X-ray observations from other telescopes.

GROND started its work in 2007 in the optical and near-infrared energy bands. It is a detector linked to the MPG/ESO telescope at La Silla Observatory. As its main results, it detected and set the distance to one of the most distant GRBs (GRB 080913, $z = 6.695$), and then it detected GRB 080916C, demonsrating that this was the most energetic GRB at the time.

VLT observes in the visible and infrared wavelengths and is composed of four independent telescopes (Antu, Kueyen, Melipal and Yepun). It is made of large-field imagers, adaptive optics corrected cameras and spectrographs, high-resolution and multi-object spectrographs, and works in energy ranges from deep ultraviolet to mid-infrared. It has an angular resolution of 0.001 arcsec when the four telescopes are working together. As a main result, together with the GROND telescope, VLT enabled the analysis of the afterglow phase of the furthest GRB at the time (GRB 080913).

1.4 The historical background of SNe

Novae and SNe have been observed for thousands of years, but their systematic study only began in 1885, when Hartwig observed a 'Nova' in the centre of the Andromeda galaxy. This object had been visible for 18 months. After 4 years it was clear that the object discovered by Hartwig was 1000 times more luminous than a typical nova. Another similar event was observed in 1895, but we had to wait until 1934 to have a clear distinction between novae and SNe. From then, thanks to Zwicky, in 22 years 54 SNe were discovered. In 1939 it was evident that there was an apparent uniformity in the intrinsic brightness at the maximum luminosity and after 15 days from the maximum. This circumstance suggested that SNe could be used as distance indicators. But even at the time, the question of the uniformity and homogeneity of the properties presented difficulties, because in 1940, SN 1940c was observed, an event with spectral properties very different to the other SNe already observed. So another distinction was necessary and SNe were classified into type I if they resembled SN 1937c or type II if they were similar to SN 1940c.

1.4.1 Classification of SNe

At the beginning of the 1940s, R Minkowski noted the presence of two main classes of SNe, characterized by the presence or absence of hydrogen in the spectrum. If these lines were observed, the SNe were called type I, otherwise they were SNe of type II. SNe I are observed in every type of galaxy and are associated with stars of population II, namely old with small mass values. This simple classification remained unchanged until the middle of the 1980s, when it became clear that SNe type I were further divided into two different types, called Ia and Ib (Uomoto and Kirshner 1985). Currently the classification is based on the properties of the observed spectrum. SNe I lack the hydrogen Balmer series at maximum luminosity, while SNe II present these lines. SNe type I are further classified into SNe Ia, SNe Ib and SNe Ic if they have lines of silicon ionized (SiII at 615.0 nm), helium lines not ionized (HeI at 587.6 nm) or if they have neither of the features, respectively. SNe type II are classified, instead, into SNe IIP or SNe IIL if the light curve presents a plateau phase or a linear decreasing trend, respectively. This class of SNe presents lines broadened through the high velocity of the ejecta. There are also sometimes SNe with very narrow lines, indicated with SNe IIn, where n means narrow. SNe Ia are produced by thermonuclear explosions of white dwarfs with a prevalence of Si or by accretion in a binary system with a more massive companion star. From a photometric point of view, they present quite a uniform light curve and a correlation between the peak luminosity and the wideness of the light curve, called the Phillips correlation (Phillips 1993). These properties are used for cosmological purposes. From the moment of explosion an SN Ia takes 15–20 days to reach maximum luminosity. The dispersion around the maximum is quite small (0.13 mag). After the maximum, the light curve of the SN shows a fast decline (2–3 order of magnitude in 30 days), followed by a slow decline in logarithmic scale (0.015 mag for day). SNe Ib are in the early stages similar to SNe Ia: the light curve has the same shape, but the magnitude at the maximum is usually smaller by 1.5 mag. The final stages are different and they do not follow the Phillips law. This is the reason why SNe Ib cannot be used as standard candles as SNe Ia are, although recent studies by Cano (2014) showed some pieces of evidence for the use of this type of SN as a standardizable candle.

References

Abbott B P *et al* 2016 Observation of gravitational waves from a binary black hole merger *Phys. Rev. Lett.* **116** 061102

Abdo A A *et al* 2009 Fermi observations of high-energy gamma-ray emission from GRB 080916C *Science* **323** 1688

Agrawal P C 2005 *Astrosat*: the first Indian astronomy satellite with multiwavelength capability, *Proc. Int. Cosmic Ray Conf.* **vol 4**, *Pune, India 3–10 August* (Mumbai: Tata Institute of Fundamental Research) p 171

Amati L 2006 The $E_{p,i}$–E_{iso} correlation in gamma-ray bursts: updated observational status, re-analysis and main implications *Mon. Not. R. Astron. Soc.* **372** 233–45

Baade W and Zwicky F 1934 On super-novae *Proc. Natl Acad. Sci.* **20** 254–9

Band D *et al* 1993 BATSE observations of gamma-ray burst spectra. I—Spectral diversity *Astrophys. J.* **413** 281–92

Barraud C *et al* 2003 Spectral analysis of 35 GRBs/XRFs observed with *HETE-2*/FREGATE *Astron. Astrophys.* **400** 1021–30

Barthelmy S D *et al* 2005 An origin for short γ-ray bursts unassociated with current star formation *Nature* **438** 994–6

Bassani L, Malaguti G, Jourdain E, Roques J P and Johnson W N 1995 Detection of soft gamma-ray emission from the Seyfert 2 galaxy NGC 4507 by the OSSE telescope *Astrophys. J.* **444** L73–6

Bégué D and Pe'er A 2015 Poynting-flux-dominated jets challenged by their photospheric emission *Astrophys. J.* **802** 134–41

Berger E 2014 Short-duration gamma-ray bursts *Annu. Rev. Astron. Astrophys.* **52** 43–105

Boella G, Butler R C, Perola G C, Piro L, Scarsi L and Bleeker J A M 1997 *BeppoSAX*, the wide band mission for X-ray astronomy *Astron. Astrophys. Suppl. Ser.* **122** 299

Briggs M S, Paciesas W S, Pendleton G N, Meegan C A, Fishman G J, Horack J M, Brock M N, Kouveliotou C, Hartmann D H and Hakkila J 1996 BATSE observations of the large-scale isotropy of gamma-ray bursts *Astrophys. J.* **459** 40

Cannizzo J K and Gehrels N 2009 A new paradigm for gamma-ray bursts: long-term accretion rate modulation by an external accretion disk *Astrophys. J.* **700** 1047–58

Cannizzo J K, Troja E and Gehrels N 2011 Fall-back disks in long and short gamma-ray bursts *Astrophys. J.* **734** 35

Cano Z 2014 Gamma-ray burst supernovae as standardizable candles *Astrophys. J.* **794** 121

Carson J 2007 GLAST: physics goals and instrument status *J. Phys. Conf. Ser.* **60** 115–8

Cavallo G and Rees M J 1978 A qualitative study of cosmic fireballs and gamma-ray bursts *Mon. Not. R. Astron. Soc.* **183** 359–65

Chincarini G *et al* 2007 The first survey of X-ray flares from gamma-ray bursts observed by *Swift*: temporal properties and morphology *Astrophys. J.* **671** 1903–20

Connaughton V *et al* 2016 *Fermi* GBM observations of LIGO gravitational-wave event GW150914 *Astrophys. J.* **826** L6

Costa E *et al* 1997 Discovery of an X-ray afterglow associated with the γ-ray burst of 28 February 1997 *Nature* **387** 783–5

Dainotti M G and Del Vecchio R 2017 Gamma-ray burst afterglow and prompt-afterglow relations: an overview *New Astron. Rev.* **77** 23–61

Dainotti M G, Willingale R, Capozziello S, Fabrizio Cardone V and Ostrowski M 2010 Discovery of a tight correlation for gamma-ray burst afterglows with 'canonical' light curves *Astrophys. J.* **722** L215–9

Dainotti M G, Del Vecchio R and Tarnopolski M 2018 Gamma-ray burst prompt correlations *Adv. Astron.* **2018** 4969503

D'Alessio V and Piro L 2005 General properties of X-ray riches and X-ray flashes in comparison with gamma-ray bursts *Nuovo Cim.* C **28** 497

Dall'Osso S, Stratta G, Guetta D, Covino S, De Cesare G and Stella L 2011 Gamma-ray bursts afterglows with energy injection from a spinning down neutron star *Astron. Astrophys.* **526** A121

Dekker H, D'Odorico S, Kaufer A, Delabre B and Kotzlowski H 2000 Design, construction, and performance of UVES, the echelle spectrograph for the UT2 Kueyen Telescope at the ESO

Paranal Observatory *Optical and IR Telescope Instrumentation and Detectors, Proc. SPIE* **vol 4008** ed Iye M and Moorwood A F pp 534–45

Della Valle M, Malesani D, Benetti S, Chincarini G, Stella L and Tagliaferri G 2006 Supernova 2005nc and GRB 050525A *IAU Circ.* **8696** 1

Duffell P C and MacFadyen A I 2015 From engine to afterglow: collapsars naturally produce top-heavy jets and early-time plateaus in gamma-ray burst afterglows *Astrophys. J.* **806** 205

Efron B and Petrosian V 1995 Testing isotropy versus clustering of gamma-ray bursts *Astrophys. J.* **449** 216

Eichler D, Livio M, Piran T and Schramm D N 1989 Nucleosynthesis, neutrino bursts and gamma-rays from coalescing neutron stars *Nature* **340** 126–8

Evans P A *et al* 2009 Methods and results of an automatic analysis of a complete sample of *Swift*-XRT observations of GRBs *Mon. Not. R. Astron. Soc.* **397** 1177–201

Evans P A *et al* 2016 *Swift* follow-up of the gravitational wave source GW150914 *Mon. Not. R. Astron. Soc.* **460** L40–4

Fishman G J and Meegan C A 1995 Gamma-ray bursts *Annu. Rev. Astron. Astrophys.* **33** 415–58

Fox D B *et al* 2005 The afterglow of GRB 050709 and the nature of the short-hard γ-ray bursts *Nature* **437** 845–50

Fruchter A 2000 Gamma-ray bursts and their host environments *HST Proposal* ID 8588

Fynbo J P U *et al* 2006 No supernovae associated with two long-duration γ-ray bursts *Nature* **444** 1047–9

Gehrels N and Razzaque S 2013 Gamma-ray bursts in the Swift–Fermi era *Front. Phys.* **8** 661–78

Gehrels N *et al* 2004 The *Swift* gamma-ray burst mission *Astrophys. J.* **611** 1005–20

Golkhou V Z and Butler N 2014 Uncovering the intrinsic variability of gamma-ray bursts *Astrophys. J.* **787** 90–9

Golkhou V Z, Butler N, Littlejohns J and Owen M 2015 The energy dependence of GRB minimum variability timescales *Astrophys. J.* **811** 93–104

González M M, Dingus B L, Kaneko Y, Preece R D, Dermer C D and Briggs M S 2003 A γ-ray burst with a high-energy spectral component inconsistent with the synchrotron shock model *Nature* **424** 749–51

Gotz D 2013 *INTEGRAL* results on gamma-ray bursts, arXiv: 1302.4847

Greiner J *et al* 2008 GROND a 7-channel imager *Publ. Astron. Soc. Pac.* **120** 405

Greiner J *et al* 2015 A very luminous magnetar-powered supernova associated with an ultra-long γ-ray burst *Nature* **523** 189–92

Groot P J, Galama T J, van Paradijs J and Kouveliotou C 1997 Optical afterglow of a gamma-ray burst: GRB 970228 *IEEE Spectr.* **14** 8–11

Guetta D 2006 Short GRBs: rates and luminosity function implications *Nuovo Cim.* **B 121** 1061–6

Hailey C J *et al* 2010 The nuclear spectroscopic telescope array (*NuSTAR*): optics overview and current status *Proc. SPIE* **7732** 77320T

Heise J, Zand J I, Kippen R M and Woods P M 2001 X-ray flashes and X-ray rich gamma ray bursts *Gamma-ray Bursts in the Afterglow* ed E Costa, F Frontera and J Hjorth (Berlin: Springer)

Hjorth J *et al* 2003 A very energetic supernova associated with the γ-ray burst of 29 March 2003 *Nature* **423** 847–50

Horváth I 1998 A third class of gamma-ray bursts? *Astrophys. J.* **508** 757–9

Horváth I 2002 A further study of the BATSE gamma-ray burst duration distribution *Astron. Astrophys.* **392** 791–3

Horváth I 2009 Classification of *BeppoSAX*'s gamma-ray bursts *Astrophys. Space Sci.* **323** 83–6

Horváth I, Balázs L G, Bagoly Z, Ryde F and Mészáros A 2006 A new definition of the intermediate group of gamma-ray bursts *Astron. Astrophys.* **447** 23–30

Horváth I, Balázs L G, Bagoly Z and Veres P 2008 Classification of *Swift*'s gamma-ray bursts *Astron. Astrophys.* **489** L1–4

Horváth I, Bagoly Z, Balázs L G, de Ugarte Postigo A, Veres P and Mészáros A 2010 Detailed classification of *Swift*'s gamma-ray bursts *Astrophys. J.* **713** 552–7

Huja D, Mészáros A and Řípa J 2009 A comparison of the gamma-ray bursts detected by BATSE and *Swift Astron. Astrophys.* **504** 67–71

Jakobsson P, Hjorth J, Fynbo J P U, Watson D, Pedersen K, Björnsson G and Gorosabel J 2004 *Swift* identification of dark gamma-ray bursts *Astrophys. J.* **617** L21–4

Jansen F *et al* 2001 XMM-Newton observatory. I. The spacecraft and operations *Astron. Astrophys.* **365** L1–6

Kaneko Y *et al* 2007 Prompt and afterglow emission properties of gamma-ray bursts with spectroscopically identified supernovae *Astrophys. J.* **654** 385–402

Kippen R M, Woods P M, Heise J, Zand J I, Preece R D and Briggs M S 2001 BATSE observations of fast X-ray transients detected by *BeppoSAX*-WFC *Gamma-ray Bursts in the Afterglow Era* ed E Costa, E Frontera and J Hjorth (Berlin: Springer)

Klebesadel R W, Strong I B and Olson R A 1973 Observations of gamma-ray bursts of cosmic origin *Astrophys. J.* **182** L85

Koen C and Bere A 2012 On multiple classes of gamma-ray bursts, as deduced from autocorrelation functions or bivariate duration/hardness ratio distributions *Mon. Not. R. Astron. Soc.* **420** 405–15

Kouveliotou C, Meegan C A, Fishman G J, Bhat N P, Briggs M S, Koshut T M, Paciesas W S and Pendleton G N 1993 Identification of two classes of gamma-ray bursts *Astrophys. J.* **413** L101–4

Kumar P and Zhang B 2015 The physics of gamma-ray bursts and relativistic jets *Phys. Rep.* **561** 1–109

Kumar P, Narayan R and Johnson J L 2008 Properties of gamma-ray burst progenitor stars *Science* **321** 376

Levan A J *et al* 2014 A new population of ultra-long duration gamma-ray bursts *Astrophys. J.* **781** 13

Li X, Zhang F-W, Yuan Q, Jin Z-P, Fan Y-Z, Liu S-M and Wei D-M 2016 Implications of the tentative association between GW150914 and a *Fermi*-GBM transient *Astrophys. J.* **827** L16

Liang E and Zhang B 2005 Model-independent multivariable gamma-ray burst luminosity indicator and its possible cosmological implications *Astrophys. J.* **633** 611–23

Liu R-Y, Wang X-Y and Wu X-F 2013 Interpretation of the unprecedentedly long-lived high-energy emission of GRB 130427A *Astrophys. J.* **773** L20

Loeb A 2016 Electromagnetic counterparts to black hole mergers detected by LIGO *Astrophys. J.* **819** L21

Longo F *et al* 2007 The *AGILE* mission and gamma-ray bursts *AIP Conf. Ser.* **906** 147–55

Malesani D *et al* 2004 SN 2003lw and GRB 031203: a bright supernova for a faint gamma-ray burst *Astrophys. J.* **609** L5–8

Mazets E P *et al* 1981 Catalog of cosmic gamma-ray bursts from the KONUS experiment data. I *Astrophys. Space Sci.* **80** 3–83

Meegan C A, Fishman G J, Wilson R B, Horack J M, Brock M N, Paciesas W S, Pendleton G N and Kouveliotou C 1992 Spatial distribution of gamma-ray bursts observed by BATSE *Nature* **355** 143–5

Meegan C A *et al* 1998 The 4B BATSE gamma-ray burst catalog *AIP Conf. Proc.* **428** 3–9

Mészáros P 1998 Theoretical models of gamma-ray bursts *AIP Conf. Proc.* **428** 647–56

Mészáros P 2006 Gamma-ray bursts *Rep. Prog. Phys.* **69** 2259–321

Mészáros P and Rees M J 2015 Gamma-ray bursts *General Relativity and Gravitation: A Centennial Perspective*ed ed A Ashtekar, B Berger, J Isenberg and M A H MacCallum (Cambridge: Cambridge University Press) 148–61

Metzger B D, Giannios D, Thompson T A, Bucciantini N and Quataert E 2011 The protomagnetar model for gamma-ray bursts *Mon. Not. R. Astron. Soc.* **413** 2031–56

Metzger M R, Djorgovski S G, Kulkarni S R, Steidel C C, Adelberger K L, Frail D A, Costa E and Frontera F 1997 Spectral constraints on the redshift of the optical counterpart to the γ-ray burst of 8 May 1997 *Nature* **387** 878–80

Morsony B J, Workman J C and Ryan D M 2016 Modeling the afterglow of the possible *Fermi*-GBM event associated with GW150914 *Astrophys. J.* **825** L24

Mukherjee S, Feigelson E D, Jogesh Babu G, Murtagh F, Fraley C and Raftery A 1998 Three types of gamma-ray bursts *Astrophys. J.* **508** 314–27

Nakar E 2007 Short-hard gamma-ray bursts *Phys. Rep.* **442** 166–236

Napier P J, Thompson A R and Ekers R D 1983 The Very Large Array—design and performance of a modern synthesis radio telescope *IEEE Proc.* **71** 1295–320

Narayan R, Paczynski B and Piran T 1992 Gamma-ray bursts as the death throes of massive binary stars *Astrophys. J.* **395** L83–6

Narayana Bhat P *et al* 2016 The third *Fermi* GBM gamma-ray burst catalog: the first six years *Astrophys. J. Suppl. Ser.* **223** 28

Nardini M, Ghisellini G, Ghirlanda G, Tavecchio F, Firmani C and Lazzati D 2006 Clustering of the optical-afterglow luminosities of long gamma-ray bursts *Astron. Astrophys.* **451** 821–33

Norris J P and Bonnell J T 2006 Short gamma-ray bursts with extended emission *Astrophys. J.* **643** 266–75

Norris J P, Nemiroff R J, Bonnell J T, Scargle J D, Kouveliotou C, Paciesas W S, Meegan C A and Fishman G J 1996 Attributes of pulses in long bright gamma-ray bursts *Astrophys. J.* **459** 393

Nousek J A *et al* 2006 Evidence for a canonical gamma-ray burst afterglow light curve in the *Swift* XRT data *Astrophys. J.* **642** 389–400

O'Brien P T *et al* 2006 The early X-ray emission from GRBs *Astrophys. J.* **647** 1213–37

Paczynski B 1991a Cosmological gamma-ray bursts *Acta Astron.* **41** 257–67

Paczynski B 1991b On the galactic origin of gamma-ray bursts *Acta Astron.* **41** 157–66

Panaitescu A, Vestrand W T and Woźniak P 2013 An external-shock model for gamma-ray burst afterglow 130427A *Mon. Not. R. Astron. Soc.* **436** 3106–11

Pe'er A 2015 Physics of gamma-ray bursts prompt emission *Adv. Astron.* **2015** 907321

Perna R, Lazzati D and Giacomazzo B 2016 Short gamma-ray bursts from the merger of two black holes *Astrophys. J.* **821** L18

Perley D A *et al* 2016 The *Swift* GRB host galaxy legacy survey. II. Rest-frame near-IR luminosity distribution and evidence for a near-solar metallicity threshold *Astrophys. J.* **817** 8

Phillips M M 1993 The absolute magnitudes of type IA supernovae *Astrophys. J.* **413** L105–8

Piran T 2004 The physics of gamma-ray bursts *Rev. Mod. Phys.* **76** 1143–210

Piro L *et al* 2000 Observation of X-ray lines from a gamma-ray burst (GRB991216): evidence of moving ejecta from the progenitor *Science* **290** 955–8

Qin Y-P, Xie G-Z, Xue S-J, Liang E-W, Zheng X-T and Mei D-C 2000 The hardness–duration correlation in the two classes of gamma-ray bursts *Publ. Astron. Soc. Jpn.* **52** 759

Rea N, Gullón M, Pons J A, Perna R, Dainotti M G, Miralles J A and Torres D F 2015 Constraining the GRB-magnetar model by means of the galactic pulsar population *Astrophys. J.* **813** 92

Reeves J N *et al* 2002 The signature of supernova ejecta in the X-ray afterglow of the γ-ray burst 011211 *Nature* **416** 512–5

Rhoads J E 1997 How to tell a jet from a balloon: a proposed test for beaming in gamma-ray bursts *Astrophys. J.* **487** L1–4

Řípa J and Shafieloo A 2017 Testing the isotropic universe using the gamma-ray burst data of *Fermi*/GBM *Astrophys. J.* **851** 15

Řípa J, Mészáros A, Wigger C, Huja D, Hudec R and Hajdas W 2009 Search for gamma-ray burst classes with the *RHESSI* satellite *Astron. Astrophys.* **498** 399–406

Rowlinson A and O'Brien P 2012 Energy injection in short GRBs and the role of magnetars *Gamma-Ray Bursts 2012 Conference (GRB 2012)*

Rowlinson A, Gompertz B P, Dainotti M, O'Brien P T, Wijers R A M J, van der and Horst A J 2014 Constraining properties of GRB magnetar central engines using the observed plateau luminosity and duration correlation *Mon. Not. R. Astron. Soc.* **443** 1779–87

Ryde F, Björnsson C-I, Kaneko Y, Mészáros P, Preece R and Battelino M 2006 Gamma-ray burst spectral correlations: photospheric and injection effects *Astrophys. J.* **652** 1400–15

Sakamoto T *et al* 2005 Global characteristics of X-ray flashes and X-ray-rich gamma-ray bursts observed by *HETE-2 Astrophys. J.* **629** 311–27

Sakamoto T, Hill J E, Yamazaki R, Angelini L, Krimm H A, Sato G, Swindell S, Takami K and Osborne J P 2007 Evidence of exponential decay emission in the *Swift* gamma-ray bursts *Astrophys. J.* **669** 1115–29

Sako M, Harrison F A and Rutledge R E 2005 A search for discrete X-ray spectral features in a sample of bright γ-ray burst afterglows *Astrophys. J.* **623** 973–99

Savchenko V *et al* 2016 *INTEGRAL* upper limits on gamma-ray emission associated with the gravitational wave event GW150914 *Astrophys. J.* **820** L36

Schulze S *et al* 2014 GRB 120422A/SN 2012bz: bridging the gap between low- and high-luminosity gamma-ray bursts *Astron. Astrophys.* **566** A102

Shirasaki Y *et al* 2003 In-orbit performance of wide-field X-ray monitor on *HETE-2 Proc. SPIE* **4851** 1310–19

Sparre M *et al* 2011 Spectroscopic evidence for SN 2010ma associated with GRB 101219B *Astrophys. J.* **735** L24

Strohmayer T E, Fenimore E E, Murakami T and Yoshida A 1998 X-ray spectral characteristics of GINGA gamma-ray bursts *Astrophys. J.* **500** 873–87

Tarnopolski M 2017 Testing the anisotropy in the angular distribution of *Fermi*/GBM gamma-ray bursts *Mon. Not. R. Astron. Soc.* **472** 4819–31

Teegarden B J and Sturner S J 1999 *INTEGRAL* observations of gamma-ray bursts AAS/High Energy Astrophysics Division #4 *Bull. Am. Astron. Soc.* **31** 717

Tegmark M, Hartmann D H, Briggs M S and Meegan C A 1996 The angular power spectrum of BATSE 3B gamma-ray bursts *Astrophys. J.* **468** 214

Tiengo A, Mereghetti S, Ghisellini G, Rossi E, Ghirlanda G and Schartel N 2003 The X-ray afterglow of GRB 030329 *Astron. Astrophys.* **409** 983–7

Troja E *et al* 2017 The X-ray counterpart to the gravitational-wave event GW170817 *Nature* **551** 71–4

Uomoto A and Kirshner R P 1985 *Astron. Astrophys.* **149** L7–L9

Usov V V 1992 Millisecond pulsars with extremely strong magnetic fields as a cosmological source of gamma-ray bursts *Nature* **357** 472–4

Veres P, Bagoly Z, Horváth I, Mészáros A and Balázs L G 2010 A distinct peak-flux distribution of the third class of gamma-ray bursts: a possible signature of X-ray flashes? *Astrophys. J.* **725** 1955–64

Virgili F J *et al* 2013 GRB 091024A and the nature of ultra-long gamma-ray bursts *Astrophys. J.* **778** 54

Watson D, Reeves J N, Hjorth J, Jakobsson P and Pedersen K 2003 Delayed soft X-ray emission lines in the afterglow of GRB 030227 *Astrophys. J.* **595** L29–32

Watson D, Hjorth J, Jakobsson P, Pedersen K, Patel S and Kouveliotou C 2004 Massive star-formation rates of γ-ray burst host galaxies: an unobscured view in X-rays *Astron. Astrophys.* **425** L33–6

Wijers R A M J, Rees M J and Meszaros P 1997 Shocked by GRB 970228: the afterglow of a cosmological fireball *Mon. Not. R. Astron. Soc.* **288** L51–6

Willingale R *et al* 2007 Testing the standard fireball model of gamma-ray bursts using late X-ray afterglows measured by *Swift Astrophys. J.* **662** 1093–110

Woosley S E and Bloom J S 2006 The supernova gamma-ray burst connection *Annu. Rev. Astron. Astrophys.* **44** 507–56

Zhang B 2011 Open questions in GRB physics *C. R. Phys.* **12** 206–25

Zhang B and Mészáros P 2001 Gamma-ray burst afterglow with continuous energy injection: signature of a highly magnetized millisecond pulsar *Astrophys. J.* **552** L35–8

Zhang B *et al* 2009 Discerning the physical origins of cosmological gamma-ray bursts based on multiple observational criteria: the cases of $z = 6.7$ GRB 080913, $z = 8.2$ GRB 090423, and some short/hard GRBs *Astrophys. J.* **703** 1696–724

Zhang B-B, Liang E-W and Zhang B 2007 A comprehensive analysis of *Swift* XRT data. I. Apparent spectral evolution of gamma-ray burst X-ray tails *Astrophys. J.* **666** 1002–11

Zitouni H, Guessoum N, Azzam W J and Mochkovitch R 2015 Statistical study of observed and intrinsic durations among BATSE and *Swift*/BAT GRBs *Astrophys. Space Sci.* **357** 7

Chapter 2

GRB models

GRB observations have shown several GRB classes. In this chapter we will give a brief description of several models. One of the most popular models is the fireball model, described in section 2.2. The general problem in pin-pointing the most appropriate model is that the 'inner engine' that produces the relativistic energy flow is hidden from direct observation. However, the observed temporal structure directly reflects this 'engine's' activity, which necessarily suggests a compact internal 'engine' that produces a wind—a long energy flow (long compared to the size of the 'engine' itself)—rather than an explosive 'engine' that produces a fireball whose size is comparable to the size of the 'engine' (Piran 1999). However, not all of the energy of the relativistic shell can be converted to radiation or to thermal energy by the internal shocks responsible for the prompt emission, see figure 2.1, of the fireball model. The remaining kinetic energy will most likely dissipate via external shocks that will produce an 'afterglow' in different wavelengths. The fireball model can explain the progenitor emission mechanism regardless of whether we focus on short or long GRBs. With the discovery of afterglow (e.g. Costa *et al* 1997), the classification of long and short GRBs found physical grounds in the evidence for two distinct progenitors for LGRBs, associated with the gravitational collapse of massive stars, and the SGRBs recently confirmed to be associated with compact binary coalescences (Abbott *et al* 2017). Two GRB progenitors have been preferentially discussed in the literature: explosions of massive stars for LGRBs, as pointed out by Woosley (1993), and NS–NS or BH–NS coalescence for SGRBs, as stated by Paczynski (1986) and Eichler *et al* (1989). There are several detailed models describing the phenomenon, but the current available observations are insufficient to constrain these models. In this chapter, the most popular GRB models will be briefly described.

2.1 The compactness problem

It is known that the GRB is emitted not from the source, where the energy was initially deposited, but from a moving material, which has large Γ. For this reason,

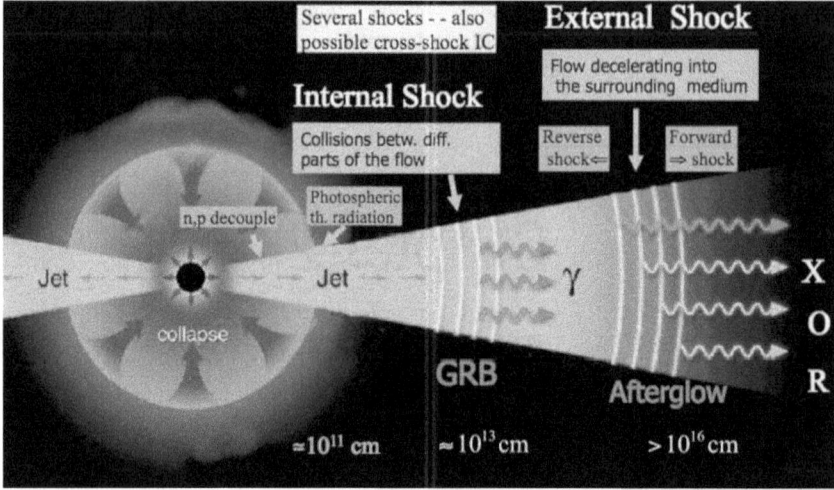

Figure 2.1. Sketch of the fireball model. Reproduced from Mészáros and Rees (2015) with permission. Copyright 2015 Cambridge University Press.

the size of the source R_0 is smaller than the size of the emitting region, R_{er}. In fact, assuming that the source is not moving relativistically, the size of the emitting region can be estimated from the timescale of the GRB temporal variability t_{var} through the relation $R_{er} \sim c t_{var}$. Given that some bursts have t_{var} of the order of milliseconds (Piran 1999), it is found that

$$R_{er} \sim 10^7 \text{ cm}, \qquad (2.1)$$

so $R_0 \lesssim 10^7$ cm. This gives rise to the so-called compactness problem (Piran 1999). In fact, the source is very compact, having the size of typical compact astrophysical objects such as NSs or BHs. Therefore, the observed energy implies very large energy density in the sources of GRBs. This leads in turn to the conclusion that the source must be optically thick to pair production via the process $\gamma\gamma \to e^+e^-$. Let us now consider the GRB with luminosity L_γ and average energy of photons ϵ_γ emitted during the time t_{var}. The optical depth[1] for pair production through the reaction $\gamma\gamma \to e^+e^-$ is

$$\tau_{\gamma\gamma} \sim n_\gamma \sigma_T R_{er} \sim \frac{3N_\gamma \sigma_T}{4\pi R_{er}^2} \sim \frac{3L_\gamma \sigma_T}{4\pi \epsilon_\gamma c^2 t_{var}}, \qquad (2.2)$$

where $\sigma_T = 6.6524 \times 10^{-25}$ cm^2 is Thomson's cross-section, and spherical geometry of the source is assumed. Taking typical parameters as $L_\gamma = 10^{52}$ erg s^{-1}, $t_{var} = 1$ ms and $\epsilon_\gamma = 1$ MeV, it is found that

$$\tau_{\gamma\gamma} \sim 10^{15}, \qquad (2.3)$$

which means that the source is indeed optically thick. Then, the only radiation one

---

[1] The optical depth is the measure of how thick a medium is for the radiation travelling through it.

would see is the emission from the photosphere, but the source is so far away that nothing would be visible and, in any case, a thermal spectrum should be observed. However, special relativistic effects can resolve this difficulty. In fact, if the emitting material moves with ultra-relativistic velocity towards us, then its proper variability timescale is

$$t_{\text{prop,var}} \sim \Gamma^2 t_{\text{var}}. \tag{2.4}$$

Thus, if we substitute t_{var} with $t_{\text{prop,var}}$, the optical depth can be smaller than unity for $\Gamma \geqslant 100$. Thus, the problem of compactness is solved when $\Gamma \geqslant 100$. In fact, similar assumptions are at the basis of the indirect methods allowing estimation of the Lorentz factors of GRBs (Lithwick and Sari 2001).

Therefore, the source expands from a very compact region, almost reaching the speed of light. The bulk of radiation is emitted far from the region of formation of the plasma, when it becomes transparent to photons, trapped initially inside by the huge optical depth. The other side of the compactness problem is that the optical depth must exist in the beginning, when the plasma is just formed. Therefore, intense interactions between electrons, positrons and photons take place in the plasma. Even if initially the plasma is composed of only photons, or only pairs, creation or annihilation of pairs soon redistributes energy between particles so that the final state will be a mixture of pairs and photons. Usually, when one deals with stationary sources in astrophysics, there is enough time for an initial state of some sort of thermal equilibrium to be achieved. In contrast, for GRBs with the timescale of expansion given by $t_{\text{var}} = 1$ ms, it is not at all clear if there is sufficient time for any type of equilibrium to be reached. Regarding this aspect, there is still a vivid debate in the literature. Some authors assumed thermal equilibrium as the initial state prior to expansion (Goodman 1986), while others did not (Cavallo and Rees 1978). This fact, together with energy assumptions, suggests that the origin of GRBs may be the gravitational collapse or mergers of compact objects such as BHs, NSs or white dwarfs (Piran 2004, Mészáros 2006, Zhang 2007, Nakar 2007).

2.2 The fireball model

The GRB non-thermal observed spectrum indicates that the sources must be optically thin, instead a wide optical depth is found (Piran 2004). This compactness problem (Ruderman and Cheng 1988), as already explained in section 2.1, has been bypassed assuming that the emitting matter moves relativistically toward the observer (Piran 1999). In this case, in fact, the optical depth decreases by $\Gamma^{(4+2\alpha)}$, where α is the high-energy spectral index, so when $\Gamma \sim 10^2$ this problem is solved.

The kinetic energy of particles at relativistic energies is the best source for a relativistic jet.

A model proposed to accelerate particles to relativistic velocities is the fireball model (Goodman 1986, Paczynski 1986) (see figure 2.1). A huge amount of gamma-ray photons should be ejected and thus the creation of $e^{\pm}$ pairs can generate a thick fireball. A thick plasma with initial energy larger than its rest mass is what defines a fireball (Piran 1999). Shemi and Piran (1990) and Paczynski (1990) took into account the effect of a baryonic load, showing that the ultimate outcome will be the transfer of all the energy of the fireball to the kinetic energy of the baryons.

In a pure radiation fireball the photons at energy E_1 may come into contact with photons at smaller energy E_2, creating electron–positron pairs through $\gamma\gamma \to e^+e^-$ if $\sqrt{E_1E_2} > m_ec^2$. Due to the opacity the radiation remains confined. This plasma is assumed to be a perfect fluid and is described by the stress–energy tensor $T^{\mu\nu}$ with pressure p, energy density ϵ and equation of state $p = \epsilon/3$. In addition to radiation and $e^{\pm}$ pairs, baryonic matter can also be present in astrophysical fireballs, either in the original jet or in the external medium (Piran 1999). The evolution of the fireball can be affected in two ways by these baryons. The thickness grows due to the electrons bound to this matter, while the accelerated baryons transform the radiation energy into bulk kinetic energy.

The dynamics of the fireball might be split in two stages: the radiation dominated phase and the matter dominated phase (Piran 1999). During the latter stage the dynamical effect of the radiation on the motion is not so crucial and thus no relevant radial acceleration is produced. Thus, Γ remains constant on streamlines. There is a specific time during the expansion in which the fireball will become optically thin. After this point the radiation and the baryons do not have the same velocity and the radiation pressure becomes negligible. Any remaining radiation will now escape freely and the baryon shells will coast with their own individual velocities.

2.2.1 Energy conversion

In the standard fireball model the energy transport occurs through the kinetic energy of a shell composed of relativistic particles. The energy transformation can happen in two ways (Piran 1999): through external shocks (Rees and Meszaros 1992) or through internal shocks (Rees and Meszaros 1994).

Internal shocks are believed to describe the observed temporal structure of GRBs (Piran 1999). Occurring at $\sim 10^{15}$ cm from the progenitor, 20% of the jet kinetic energy is transformed into thermal energy (Piran 1999) and almost 50% of the shell's energy (Kobayashi et al 1997, Katz 1997) is taken out through internal shocks.

Sari and Piran (1997) claimed that in the case that the external medium is interstellar medium (ISM), collisionless shocks occur and the relativistic shell is dispersed by external shocks. This implies a further burst which explains the afterglow phase (Piran 1999). Therefore, the standard fireball model is characterized by an 'internal–external' scenario (Sari and Piran 1997) in which the prompt GRB phase is generated by internal shocks, while the afterglow phase is caused by external shocks.

In the case that a cold shell (with internal energy negligible with respect to the rest mass) approaches another cold shell or passes through the cold ISM, two kind of shocks are produced: a shock going forward in the ISM or in the external shell, and a reverse shock (RS) moving backwards in the inner shell. The Lorentz factor Γ and f, which is the ratio between the particle densities, define the structure of the jet.

There are three interesting cases (Piran 1999):

- Ultra-relativistic shock ($\Gamma \gg 1$ and $f > \Gamma^2$) occurs at the beginning of an external shock or during the very late propagation of the internal shock, when only one shock is present. Given that in this case the RS is non-relativistic or mildly relativistic (Piran 1999), this pattern is called Newtonian. It was claimed that the energy conversion happens in the forward shock (FS) (Piran 1999).

- The density ratio diminishes during the shell diffusion and $f < \Gamma^2$; in addition the FS and the RS are relativistic.
- Internal shocks are characterized by $f \approx 1$, namely the density is analogous for both shells, and $\Gamma \sim$ a few units, describing the relative motion of the shells. Both shocks are mildly relativistic and the factor Γ of the two shells affects their power.

2.2.2 Light curves within the fireball model

After having described the theoretical process of the internal shock model that produces the prompt and afterglow phases, it is relevant to investigate how the observed behaviour of the GRB light curves has been interpreted (Piran 1999, Mészáros 2002, Nakar and Piran 2002, Piran 2004).

The prompt phase within the internal shock
According to the fireball model the pulse duration and the time distance between the pulses depend on the same quantity, namely the time interval between the shells emitted by the inner engine (see e.g. Piran 2004).

This argument has been investigated using various numerical simulations (Kobayashi *et al* 1997, Daigne and Mochkovitch 1998, Panaitescu *et al* 1999), which argue that for internal shocks the light curve reproduces the temporal activity of the central engine.

To determine the time interval Δt_a between the pulses, multiple collisions should be taken into account. The angular time without cooling time defines the pulse width: $\delta t_a = R_s/(2c\Gamma_s^2)$, where Γ_s represents the value of the Lorentz factor for the radiating area. Here we mention that the synchrotron emission and the photospheric emission, described later, belong to the emission mechanism related to the prompt emission. In particular, the peak of the spectrum, E_{peak}, which is commonly used in many correlations described in this book, depends on various parameters (Zhang and Mészáros 2002). The photospheric model is an important ingredient of the fireball model. Below we give a more detailed explanation, especially in view of the fact that several prompt GRB relations, such as the Amati, the Yonetoku and the Ghirlanda relations, are poorly explained by this model. As has already been discussed for the compactness problem, in the inner part of the outflow the optical depth is too high ($\tau_{\gamma\gamma} \sim 10^{15}$). Therefore, this region is defined as the most internal part of the outflow from which the radiation can be observed. To explain the prompt phase, a signature from this region should be present in the GRB spectra, but instead the observed spectrum is non-thermal. This situation is well summarized by Pe'er (2015), where the study of this inconsistency leads to the discovery of two aspects: the spectrum from the photosphere is difficult to reconstruct, and the photosphere can be affected by energy dissipation. The first aspect was presented by Pe'er (2008) and Beloborodov (2011). From their analysis it was concluded that during the emission the comoving energy is different for each photospheric photon, causing the widening of the photospheric spectrum. In this way the spectrum becomes non-thermal. Alternatively, the second aspect, investigated by Giannios (2006) and Lazzati and Begelman (2010), states that a fraction of the kinetic energy during the prompt emission is dissipated in regions of moderate or low optical depth. The electrons are accelerated at ultra-relativistic velocity if they receive a broad fraction of this dissipated energy. Subsequently, these particles release energy through synchrotron radiation and scatter the thermal photons to higher energy, generating non-thermal spectra. In addition, the photons do not have enough time to thermalize before separating from the plasma, providing a non-thermal photospheric spectrum. For a recent review of the photospheric model, see Beloborodov and Mészáros (2017).

To have a more complete picture, sometimes the photospheric model and the external shock model are employed together. The GRB photosphere and external shock models assume that the X-ray emission is subject to the jet photospheric emission, while the synchrotron radiation from the external shocks in the jet controls the optical emission.

The afterglow phase within the external shocks
In the fireball model, the external medium making the relativistic ejecta slower gives rise to the afterglow phase (Mészáros and Rees 1997, Piran 1999). The propagation of a relativistic shell into the ISM was developed by Blandford and McKee (1976). There are two kind of ISM we can consider: the constant medium and the wind medium. When the material ejected is decelerated by the circumburst matter, the slower material catches up and generates 'refreshed shocks' (Rees and Mészáros 1998, Kumar and Piran 2000, Sari and Mészáros 2000). The extra injection of energy from these 'refreshed shocks' modifies the blast wave dynamics (Rees and Mészáros 1998, Sari and Mészáros 2000), causing a slower decay in the light curve (Piran 2004). As presented in figure 2.2, in this model the external medium structure plays an important role. Panaitescu and Kumar (2000) analysed the differences between a wind-like medium and a constant density medium. The equation for the constant density medium is the following:

$$n(r) = A \times r^{-s}, \tag{2.5}$$

where $s = 0$ for a constant density medium and $s = 2$ for a wind-like medium. Panaitescu and Kumar (2000) examined the emission behaviour in the cases in which the electrons interacting with these two kinds of media cool adiabatically or radiatively. They found a number of differences, for example in the frequency ranges characterizing the electron cooling spectra, in the plateau phases, and in the decaying parts of the afterglow phase of the light curve. However, these results do not adequately constrain the physical parameters describing GRBs. In addition to the FS going outwards, a RS diffusing into the slower material is produced when this material catches up the faster one (Kumar and Piran 2000). In addition to the refreshed shocks model, another injection model invokes a central engine (likely a

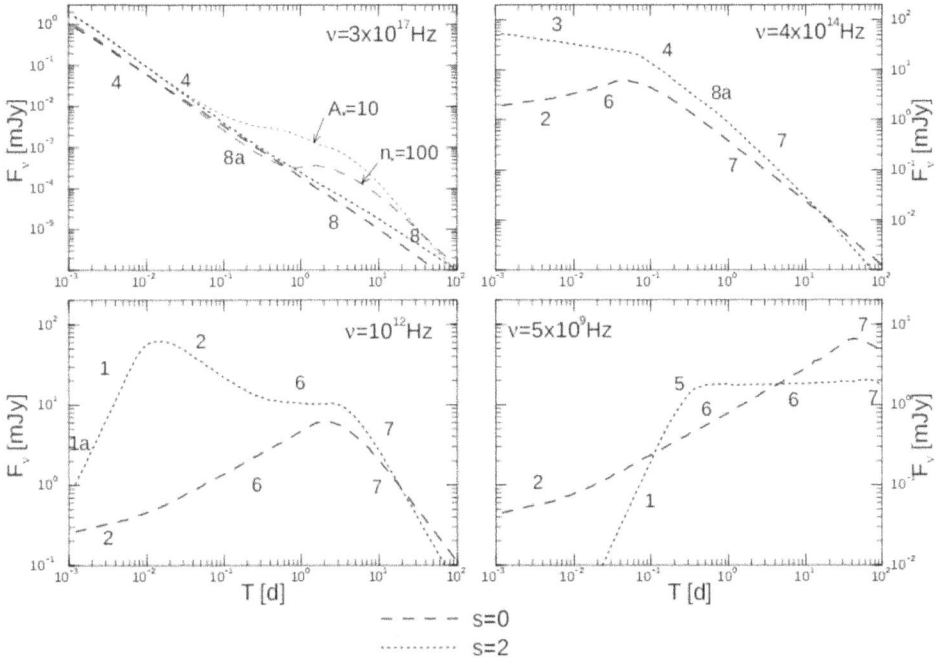

Figure 2.2. GRB afterglow light curve flux F_ν in the GRB photosphere and external shock model in four emission frequencies, see the detailed description of the presented cases in Panaitescu and Kumar (2000). The case $s = 0$ represents the constant external medium density, while the case $s = 2$ stands for the external medium formed by the stellar wind, see equation (2.5). Reproduced with permission from Panaitescu and Kumar (2000). Copyright 2000 The American Astronomical Society.

magnetar), and the two scenarios are degenerate given certain relations between the injection indices in the two models (Zhang *et al* 2006). Energy injection models, such as the magnetar, are discussed later within the afterglow correlations. Please note that the magnetar model is currently able to explain both short and long GRBs (Dall'Osso *et al* 2011, Rowlinson *et al* 2014).

2.3 The jet opening angle

Jets were initially introduced to solve the energetics problem, namely why the energy we see is much higher than the energy we should see according to the theoretical models. If the emission is not isotropic, but collimated by a narrow beaming angle $\theta_{\rm jet}$, (Mao and Yi 1994), the total energy is reduced. Within the scenario of the jet model, the emitting surface is proportional to Γ until $\Gamma \gg \theta_{\rm jet}^{-1}$. Under this condition, while the plasma expands Γ decreases and the emitting surface increases. When Γ becomes $\sim 1/\theta_{\rm jet}$, the so-called jet break in the light curve of the GRB afterglow phase can be observed (Rhoads 1999, Sari *et al* 1999). If $\Gamma^{-1} > \theta_{\rm jet}$, the radiation arrives far away from the initial jet causing the detection of a jet break for an observer looking at the original jet. A sketch of the jet opening angle with an on-axis observer, its relation to the emitting source of the progenitor, and the Γ factor is presented in figure 2.3.

An observer off-axis will detect an orphan afterglow, namely an afterglow phase without an earlier GRB. However, there is observational evidence of jets that exceed 37°, as in the case of GRB 060218. These pieces of evidence underline the fact that the fireball model may present problems in explaining these features of the observational data. As presented in Sari *et al* (1999) and Frail *et al* (2001), the jet opening angle is given by

$$
\begin{aligned}
\theta_{\text{jet}} = 0.057 \times & \left(\frac{T^*_{\text{break}}}{1 \text{ day}} \right)^{3/8} \times \left(\frac{1+z}{2} \right)^{-3/8} \\
& \times \left(\frac{E_{\text{iso}}}{10^{53} \text{ erg}} \right)^{-1/8} \times \left(\frac{\eta}{0.2} \right)^{1/8} \times \left(\frac{n}{0.1 \text{cm}^{-3}} \right)^{1/8},
\end{aligned}
\tag{2.6}
$$

where T^*_{break} is the jet break (with the * symbol used for rest-frame quantities), z is the GRB redshift, η is the jet efficiency and n is the external medium density (Frail *et al* 2001, Piran 2004). The mean value of the jet opening angle has been evaluated to be around 10° for the *Swift* measurements and 7° for the pre-*Swift* measurements (Le and Dermer 2007, Goldstein *et al* 2016). As shown by Xin *et al* (2016) and Troja *et al* (2016), the jet opening angle can be an important parameter for defining the afterglow phase. We here underline the fact that, notwithstanding that all the correlations involving E_γ are listed as prompt parameters for simplicity, they also involve afterglow observations.

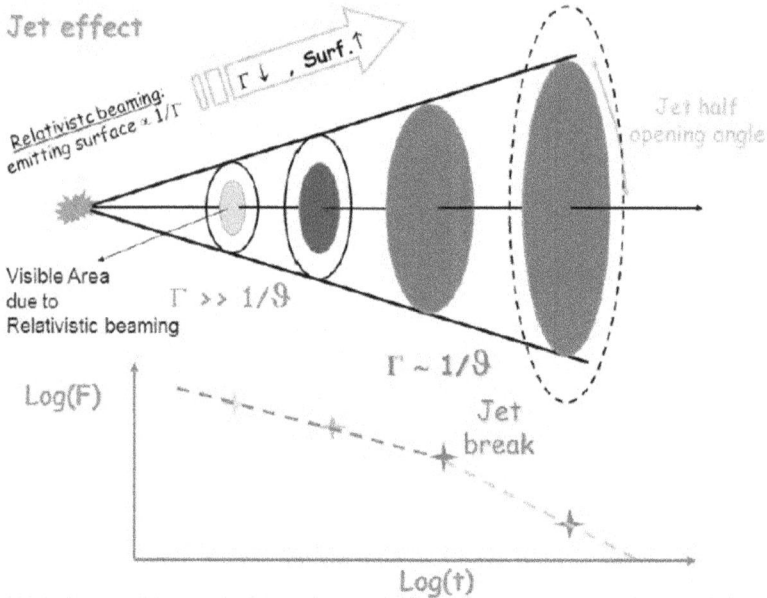

Figure 2.3. Sketch of the jet opening angle, its relation with the emitting source of the progenitor, and the Γ factor.

2.4 The central engine models

The fireball model is independent of the details of the inner engine that releases the initial energy (Mészáros 2002). However, simple assumptions on the energetics and the timescales led to the idea that the central engine producing the required energy flow could be an accretion disk. Many frameworks could lead to a BH with a massive accretion disk system. This could include mergers, such as NS–NS (Eichler *et al* 1989, Narayan *et al* 1992), NS–BH (Paczynski 1991) or NS–white dwarf (Fryer *et al* 1999b) binary systems, and models founded on 'failed SNe' or collapsars (Woosley 1993, Paczyński 1998, MacFadyen and Woosley 1999). Models for LGRBs include mechanisms such as BH accretion (Page and Thorne 1974), the rotation of a BH powering the Blandford–Znajek jet generation mechanism (Hartmann and Woosley 1995), the collapse of a massive star (MacFadyen *et al* 2001) and the magnetar model, while for SGRBs there is the magnetar model (Usov 1992), the supranova model (Vietri and Stella 1998) and the merging NSs model (Duncan and Thompson 1992, Narayan *et al* 1992). Starting with Dall'Osso *et al* (2011) the magnetar model also began to be used for explaining LGRBs. These models are described briefly in the following sections.

2.4.1 Accretion model of Page and Thorne (1974)

In the accretion model of Page and Thorne (1974), the disk sustained by fall-back material from the explosion of a massive stellar progenitor accretes onto a BH involving neutrino cooling (see figure 2.4). The energy is then released as a GRB jet. Currently, this is one of the most validated and widely used scenarios for the formation of LGRBs (Piran 2004).

In their work, Page and Thorne (1974) investigated the dynamics of the accretion disk, starting from the assumption that the BH is stationary and axially symmetric, and the accretion disk is placed in the equatorial plane of the BH. The energy flux radiated from the surface of the accretion disk is a function of the distance. In addition, Thorne (1974) analysed the evolution of the BH in the context of the accretion model. As a result, he confirmed the high efficiency of the conversion of the accreting mass into energy (30%) and constrained the value for the ratio of the spin to the BH mass suitable for the model $(0.998)^2$.

It is worth mentioning that a merging system constituting NS–NS or BH–NS also forms a disk accreting finally onto a BH. However, in this case, due to the different physical conditions of the accretion disk, an SGRB will be produced (Narayan *et al* 2001).

2.4.2 Rotating BH model

Another model for LGRBs and SGRBs was proposed to be a rotating BH in a strong magnetic field with the rotational (spin) energy extracted by the magnetic field through the Blandford–Znajek mechanism (Blandford and Znajek 1977). In

---

[2] This value is compatible with the results for a rotating BH (Kerr 1963, Newman *et al* 1965).

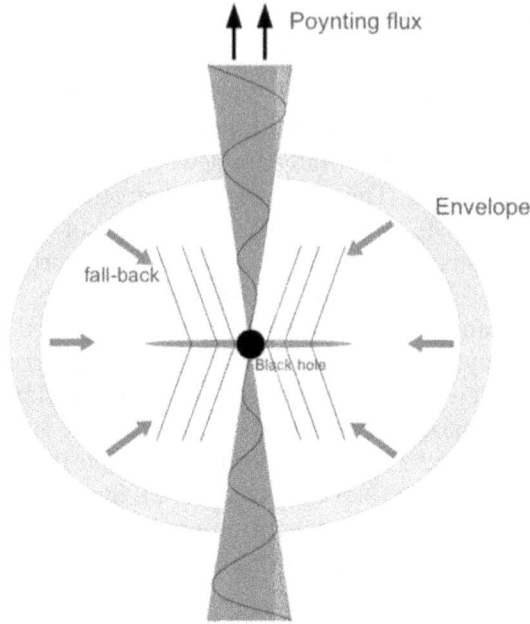

Figure 2.4. A representation of the GRB accretion model. Reproduced with permission from Wu *et al* (2013). Copyright 2013 The American Astronomical Society.

this case the power can be estimated using the following, widely used, formula by Thorne *et al* (1986):

$$P \sim 10^{49} \left(\frac{ac}{GM} \right)^2 \times \left(\frac{M}{M_\odot} \right)^2 \times \left(\frac{B}{10^{15} \, \text{G}} \right)^2 \, \text{ergs}^{-1}, \qquad (2.7)$$

where B is the magnetic field strength, a is the BH spin, M is the BH mass and $M_\odot$ represents the solar mass (1.99×10^{30} kg). This formula allows for estimation of the energy emitted from the BH for comparison to GRB observations.

Further theoretical results obtained within this model are often in agreement with those described by Hartmann and Woosley (1995). Indeed, they presented the investigation of the models describing the BH accretion and its jet. They derived the energy rate of the radiation emitted by the accretion process to be $\sim 5 \times 10^{52} \times (10 \, \dot{M}/M_\odot \text{s}^{-1})$ erg s^{-1} (with $\dot{M}$ the mass accretion rate), and the angular velocity of the BH to be $\sim 0.1 \times \left(\frac{M_{\text{disk}}}{M_\odot} \right) \times \left(\frac{M_{\text{BH}}}{5 \, M_\odot} \right)^{1/2} \times \left(\frac{10^8 \, \text{cm}}{b} \right)^{2.5} \text{s}^{-1}$, with b the distance from the disk to the BH. These values remain the best estimates for these parameters to date.

2.4.3 Magnetar model

Regarding SGRBs, a possible scenario was developed by Usov (1992) in which a strongly magnetized NS (namely a magnetar) can be an engine for GRBs (see figure 2.5). In this model a magnetar with $B \sim 10^{14-15}$ G created in a freshly collapsed massive star releases a huge amount of energy, allowing it to power the jet of an SGRB (Metzger *et al* 2011, Dall'Osso *et al* 2011, Rowlinson *et al* 2013, 2014). The magnetars involved in this process need to have a rotation period of ~1 ms, or even shorter, to fulfil the budget of total and emitted energies (Zhang and Mészáros 2001).

There are also studies in which magnetars connected to SNe and magnetars connected to LGRBs were compared, finding similarities between these two phenomena, as for example in Dall'Osso *et al* (2011), Rowlinson *et al* (2014) and Yu *et al* (2017). Indeed, the host galaxies of both GRBs and SNe seem to have a high star formation rate and low metallicity (Lunnan *et al* 2014). One should note that in the magnetar model it is often difficult to distinguish between wind-like and constant external media surrounding the GRB (Gompertz and Fruchter 2017), however, this detail can play a crucial role because different external medium densities can imply different energy emission processes. Note that as a possible explanation of LGRBs the magnetar is currently a very good candidate.

2.4.4 The collapsar

Woosley (1993) suggested that GRBs are produced through the collapse of a fast rotating Wolf–Rayet star. Paczyński (1998) concluded that GRBs are associated with the collapse of stars with $M > 30 M_\odot$. MacFadyen and Woosley (1999) investigated the propagation of a jet moving with relativistic velocities through the stellar envelope of the collapsing star and Mészáros and Rees (2001) investigated the collimation of the jet. Finally, Zhang *et al* (2003) validated further the

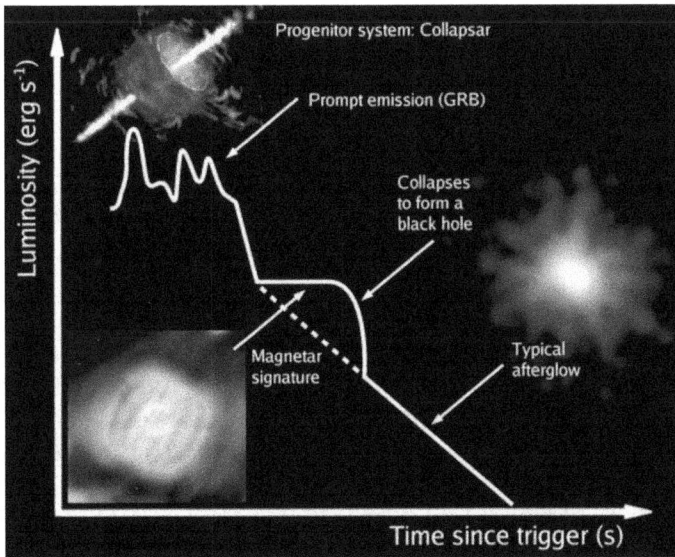

Figure 2.5. A sketch of the magnetar model. Reproduced from Rowlinson *et al* (2011). Copyright 2011 AIP Publishing.

characteristics of the jet collimation. All the aspects described below give rise to the collapsar model.

Very massive stars fuse material in their core up to the iron layer, and when they cannot produce energy by fusion they collapse, generating a BH. The central matter of the star produces an accretion disk with mass 0.1 $M_\odot$. The fall down of this material into the BH causes a couple of jets (with $\theta_{jet} < 10°$) generating a relativistic shock wave at the front (Blandford and McKee 1976). The core collapse, the accretion and the jet emission last ~10 s (MacFadyen and Woosley 1999), while the accretion onto the BH should last several dozens of seconds. The jet motion, as it breaks the envelope of the star, is influenced by its interaction with the surrounding medium. In this way the collapsar scenario attempts to model the time evolution of the prompt emission. In addition, this model accounts for a variable Γ for the production of the internal shocks (Woosley and Bloom 2006). This model is also able to predict that the activity of the central engine lasts long after the prompt phase is over (Burrows *et al* 2005). This is allowed within this model because the jet and disk are inefficient at ejecting all the material in the equatorial plane of the star before the collapse. Indeed, part of this material keeps falling back and accreting (MacFadyen *et al* 2001). If the hydrogen layer of the star is not thick, the stellar surface will be crossed by the jet. In the case the density of the stellar matter decreases, the shock accelerates, and at the surface it has a value of Γ around 100 or larger with its energy emitted as gamma-rays. A sketch of the collapsar model is presented in figure 2.6.

Evidence and problems for the collapsar view. In the collapsar model three constraints are present for a star to produce a GRB:

- the mass of the star needs to be large (at least 30 solar masses on the main sequence) to produce a BH;
- to generate a torus that is able to emit a jet, the rotation of the star has to be high;
- low metallicity of the star is required to remove its hydrogen layer, so the surface can be reached from the jets.

In order to satisfy all these requirements, it happens that core collapse SNe are more frequent than GRBs. This model is supported by the fact that LGRBs are observed in systems with high star formation, such as irregular galaxies and spiral galaxies (Pontzen *et al* 2010).

In addition, for systems at low redshift, many cases of GRBs associated with SNe Ib/c are observed. In particular, this type of SNe lacks hydrogen absorption lines. The most important GRBs associated with SNe are: GRB 980425/SN 1998bw (Galama *et al* 1998), GRB 030329/SN 2003dh (Stanek *et al* 1999, Hjorth *et al* 2003), GRB 031203/SN 2003lw (Malesani *et al* 2004), GRB 060218/SN 2006aj (Della Valle *et al* 2006), GRB 111209A/SN 2011kl (Greiner *et al* 2015), GRB 130215A/SN 2013ez, GRB 130831A/SN 2013fu (Cano *et al* 2014), GRB 130702A/SN 2013DX (Toy *et al* 2016), GRB 161219B/SN 2016jca (Cano *et al* 2017) and a few GRBs with SN bumps in their afterglow light curves.

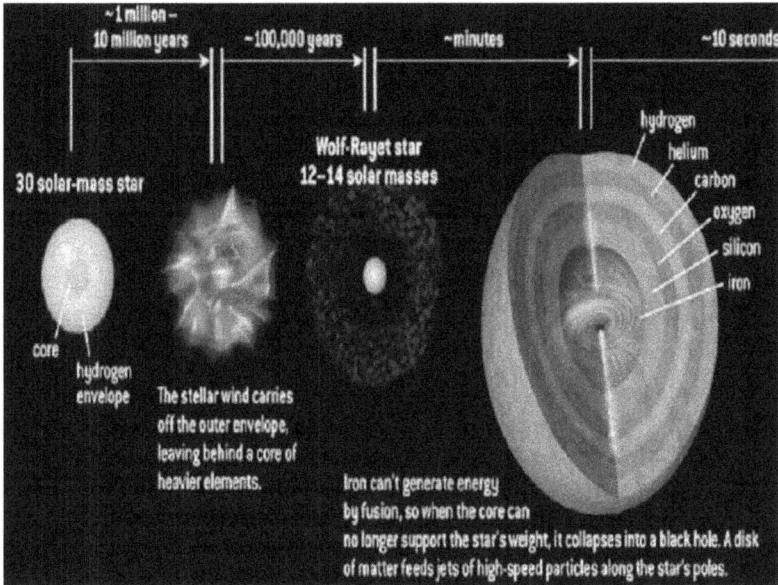

Figure 2.6. A sketch of the collapsar model summarizing the main phases from the progenitor star until the collapse. Courtesy of Dr L Vetere.

However, Fynbo *et al* (2006) pointed out that GRB 060614 and GRB 060505 do not have any associated SN, but they are observed in a high star formation region. Therefore, they concluded that, if the BH absorbs the star before the SN burst reaches the surface, the SN could not appear.

2.4.5 Merging neutron stars

Finally, for SGRBs there is the widely accredited model involving the collision of two NSs (Duncan and Thompson 1992, Narayan *et al* 1992). Within this scenario, a BH accreting the disk around the coalescent NS system can be generated (Fryer *et al* 1999a). The energy released during this process will lead to the creation of the SGRB, given the energy budget and the timescale of this phenomenon. In more detail, Narayan *et al* (1992) investigated NS–NS and BH–NS systems, taking into account different interaction mechanisms such as magnetic flares and neutrino interactions. From their model, they assumed the distance of the GRB and predicted the possible detection of GWs from these cosmological systems based on the LIGO technical specification.

On the other hand, considering other merging systems, Fryer *et al* (1999a) analysed the BH accretion phenomenon in different kinds of stellar binary systems to provide some limits for the progenitors and the formation rate of the merging binary systems. They also investigated the merging NS–helium star system and concluded that to have the highest formation rate of merging binary systems the helium core in the NS–helium star system is relevant. In fact, the presence of an NS merging with its helium companion star leads to a formation rate of merging binary systems one order of magnitude smaller. Nevertheless, the formation rate of the NS–helium star merging system is sufficient to allow the formation of an SGRB.

According to several authors, the merging NSs model seems to represent the main explanation for the origin of SGRBs (Berger 2014). Unfortunately, this model has to face two issues: the first is the occurrence of an extended emission in some SGRB light curves (Norris and Bonnell 2006), and the second is the detection of optical flares at late times (>10^5s, Gehrels *et al* 2009). Indeed, the timescales computed for this model are shorter than those required by these observational features. However, as already pointed out in chapter 1, the observations of GW170817 leave no doubt that this scenario is definitely favoured among the plausible scenarios for the creation of SGRBs.

2.5 Additional models

Several models for the GRB emission that do not focus on the central engine have also been presented in the literature. Below, for a more complete overview, several models are described. In these models the GRB emission is (a) influenced by a jet observed off-axis or (b) modified by scattering in the ISM.

2.5.1 GRB model for a jet observed off-axis

For the GRB jet observed off-axis in the afterglow phase, the viewing angle should be slightly larger than θ_{jet}. This would be the main physical explanation for the shallow plateau phase, because the viewing angle affects the emission detected shortly after the prompt phase, and the observed emission is dimmer off-axis.

As shown by Yamazaki *et al* (2003), the off-axis observation of GRB emission can lead to the detection of phenomena that look much weaker than they actually are. As pointed out by Ioka and Nakamura (2001) (see figure 2.7), this effect could allow one to interpret XRFs as LGRBs seen off-axis. Also Oates *et al* (2012) and Oates *et al* (2015) claimed that the observation of GRBs off-axis can be an explanation for the behaviour of the afterglow phase of the GRB light curve, when accompanied by a complex jet structure.

Ryan *et al* 2015, investigating a sample of 226 *Swift* GRBs through hydro-dynamic simulations, found that most of their afterglow phases were observed off-axis. In addition, they claimed that this aspect could affect the jet break behaviour in the GRB light curves.

2.5.2 Models explaining the afterglow: microphysical parameter model evolving with the Lorentz factor

To explain the differences between the GRB emission in optical and X-ray wave-lengths in the context of a modification of the fireball model with the addition of microphysical parameters (Panaitescu *et al* 2006), the physical quantities describing the GRB shock are assumed to be power-laws. Therefore, the required parameters should be: an injected energy larger than a given Γ, $E(>\Gamma) \propto \Gamma^{-e}$, the energy density

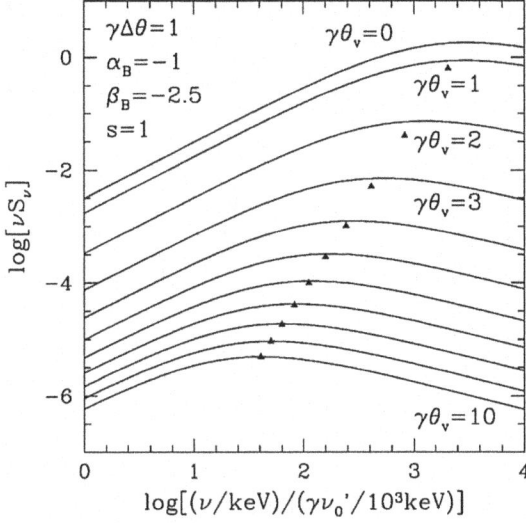

Figure 2.7. The off-axis viewing GRB jet model. Total fluence, S_ν, is plotted versus the observed frequency at different viewing angles. Reproduced with permission from Ioka and Nakamura (2001). Copyright 2001 The American Astronomical Society.

of the magnetic field $\epsilon_B \propto \Gamma^{-b}$, the energy density of the electric field $\epsilon_E \propto \Gamma^{-i}$ and the ambient medium density $n(r) \propto r^{-s}$. e, i and b are the exponents for the Lorentz factor. In this model, the difference between the temporal decay indices in X-ray ($\alpha_{X,a}$) and optical ($\alpha_{O,a}$) wavelengths is given by Urata *et al* (2007):

$$\alpha_{X,a} - \alpha_{O,a} = \frac{s}{8-2s} - \frac{1}{4} + \frac{3-s}{e+8-2s}\left[\left(\frac{s}{8-2s} - \frac{1}{4}\right)e - \frac{3}{4}b\right]. \qquad (2.8)$$

Hence, for a constant ($s = 0$) medium density, the magnetic field and the amount of energy injected are the only quantities influencing the difference between the optical and the X-ray decay rates of the GRB light curve, as claimed by Panaitescu *et al* (2006) and Urata *et al* 2007.

2.5.3 Models explaining the plateau and afterglows: GRB emission modified by scattering in the interstellar dust

The dust scattering model was developed to describe the plateau phase at the beginning of the GRB afterglow emission. This model was introduced by Shao and Dai (2007) and Shao *et al* (2008). They used the analysis carried out by Miralda-Escudé (1999). In this approach the scattering of the X-ray afterglow radiation along the GRB line of sight is due to the dust inside the parent galaxy. The authors claimed that the afterglow phase would result from the scattering of the prompt emission radiation by the dust in the medium surrounding the GRB progenitor (Shao and Dai 2007, Shao *et al* 2008).

In addition, in the framework of this model $\beta_{X,a}$ should increase, and the spectrum of the steep decay of the afterglow phase should be softer than the one of the plateau phase. However, as noted by Shen *et al* 2009, such spectral softening is not observed in the sample they analysed. In contrast, a hardening behaviour of $\beta_{X,a}$ was found, but from the analysis they could not rule out the model.

2.6 The SN Ib/c models

The position of SNe in the Galaxy can give information about their nature and the mass of the progenitor. Indeed, SNe II, Ib and Ic are mostly found in the surrounding of spiral arms and HII regions (Filippenko 1988), suggesting a mass of $8-10 \, M_\odot$ for the progenitor. Filippenko (2005) studied the progenitors of SNe II and he found that some of SNe II may originate from ejecta interacting with high-density circumstellar medium, while others originate from highly luminous and variable stars (Humphreys and Stanek 2005). From their analysis, James and Anderson (2006) claimed that the SN Ib, Ic and II fractions are proportional to the luminosity and the host galaxy star formation rate (SFR). SNe Ib and SNe Ic prefer bright host galaxies, while SNe II are most likely found in dim host galaxies. In contrast, SNe Ia are found in every kind of galaxy, with some of them favoring the spiral ones. Della Valle and Livio (1994) concluded that, due to the growing rate from early to late Hubble type, the progenitor stars for SNe Ia most likely have a mass around $4-7 \, M_\odot$ and an age of 0.1–0.5 billion years. Li *et al* (2011) investigated the luminosity functions of different SNe. They found there is no correlation between the SN luminosity function and the host galaxy dimensions. Additionally, they claimed that SNe Ib and SNe Ic appear to originate from luminous galaxies. Fink *et al* (2007) analysed a model in which a white dwarf explodes before reaching the Chandrasekhar limit, but this model looks unlikely for SNe Ia.

Instead, Ouyed and Staff (2013) developed a model in which quark-nova ejecta colliding with a white dwarf at the Chandrasekhar limit can generate a phenomenon similar to SNe Ia. The few investigated SNe Ib/c present a light curve with a rise time and fast-decline phase lasting 15–20 days and 30 days, respectively, and a decaying phase with a rate of 0.01–0.03 mag day^{-1}. SN Ib/c light curves are fuelled by the process Ni $^{56} \rightarrow$ Co $^{56} \rightarrow$ Fe^{56}, differently to those for SNe IIP. While the amount of produced nickel provides the peak of the emission, the photon diffusion indicates the width. This diffusion is constrained by the envelope mass and the expansion velocity.

The determination of the kinetic energy is derived from the Doppler effect widening the spectral lines. The decaying at late time shows that part of the gamma-rays leave the SN ejecta not thermalized, and thus we are able to estimate the amount of Ni^{56} in the SN core. Jerkstrand *et al* (2015) found that the burning silicon amount can give constraints on the progenitor and the explosion mechanism. Nomoto and collaborators, assuming a spherically symmetric emission, described SNe Ib as helium stars lacking a hydrogen layer, and SNe Ic as carbon/oxygen nuclei without a helium layer. For a recent review on this topic see Nomoto and Leung (2018).

For SNe Ib/c, as for SNe II, heavier envelope masses lead to more energetic explosions, but SNe Ib/c undergo a larger mass loss. These phenomena are called hypernovae due to their extreme energy. Despite the higher than normal energies, none of these phenomena generates high nickel masses in comparison to lower energy SNe Ib/c (see figure 2.8). This result confirms that the core physics of SNe Ib/c and SNe II could be similar. Taking into account all SNe II and SNe Ib/c, it seems that there is a continuous distribution of energies below 8 foe (1 foe $\sim 10^{51}$ erg). During the explosion, in this model SNe II can emit energies in agreement with those of SN 1997ef and SN 2002ap. Even though the process responsible for the hypernova is not clear, if SN 1997ef and SN 2002ap are explained within this model (Deng *et al* 2003), then at least one SN II (SN 1992am) can be regarded as a hypernova. To confirm that the energy distribution is continuous, more data are needed. Smartt *et al* (2002) analysed the images of Galaxy M74 before SN 2002ap. Even if they did not find any possible progenitor, they could not discard the Wolf–Rayet star scenario. They claimed that another possible description is that SN 2002ap may originate from a star in a binary system which has all the hydrogen and helium cores removed via mass transfer. Van Dyk *et al* (2003) could not conclude if

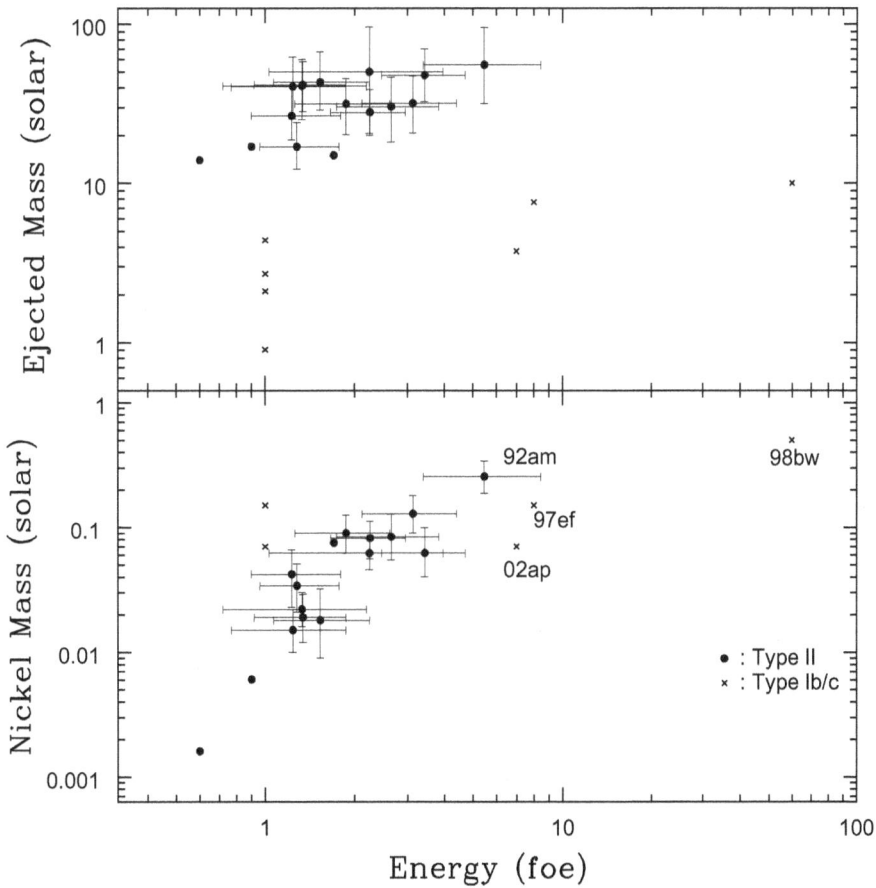

Figure 2.8. Energy of the explosion versus envelope masses and nickel masses for seven SNe Ib/c and 16 SNe II. Reproduced with permission from Hamuy (2003). Copyright 2003 The American Astronomical Society.

the Wolf–Rayet star or the massive interacting binary system are the most plausible scenarios for SN Ib/c progenitors.

References

Abbott B P *et al* 2017 Gravitational waves and gamma-rays from a binary neutron star merger: GW170817 and GRB 170817A *Astrophys. J.* **848** L13

Beloborodov A M 2011 Radiative transfer in ultrarelativistic outflows *Astrophys. J.* **737** 68

Beloborodov A M and Mészáros P 2017 Photospheric emission of gamma-ray bursts *Space Sci. Rev.* **207** 87–110

Berger E 2014 Short-duration gamma-ray bursts *Annu. Rev. Astron. Astrophys.* **52** 43–105

Blandford R D and McKee C F 1976 Fluid dynamics of relativistic blast waves *Phys. Fluids* **19** 1130–8

Blandford R D and Znajek R L 1977 Electromagnetic extraction of energy from Kerr black holes *Mon. Not. R. Astron. Soc.* **179** 433–56

Burrows D N *et al* 2005 Bright x-ray flares in gamma-ray burst afterglows *Science* **309** 1833–5

Cano Z *et al* 2014 A trio of gamma-ray burst supernovae: GRB 120729A, GRB 130215A/SN 2013ez, and GRB 130831A/SN 2013fu *Astron. Astrophys.* **568** A19

Cano Z *et al* 2017 GRB 161219B/SN 2016jca: a low-redshift gamma-ray burst supernova powered by radioactive heating *Astron. Astrophys.* **605** A107

Cavallo G and Rees M J 1978 A qualitative study of cosmic fireballs and gamma-ray bursts *Mon. Not. R. Astron. Soc.* **183** 359–65

Costa E *et al* 1997 Discovery of an x-ray afterglow associated with the γ-ray burst of 28 February 1997 *Nature* **387** 783–5

Daigne F and Mochkovitch R 1998 Gamma-ray bursts from internal shocks in a relativistic wind: temporal and spectral properties *Mon. Not. R. Astron. Soc.* **296** 275–86

Dall'Osso S, Stratta G, Guetta D, Covino S, De Cesare G and Stella L 2011 Gamma-ray bursts afterglows with energy injection from a spinning down neutron star *Astron. Astrophys.* **526** A121

Della Valle M and Livio M 1994 On the progenitors of type IA supernovae in early-type and late-type galaxies *Astrophys. J.* **423** L31–3

Della Valle M *et al* 2006 An enigmatic long-lasting γ-ray burst not accompanied by a bright supernova *Nature* **444** 1050–2

Deng J, Mazzali P A, Maeda K and Nomoto K 2003 The type Ic hypernova SN 2002ap *Nucl. Phys.* A **718** 569–71

Duncan R C and Thompson C 1992 Formation of very strongly magnetized neutron stars—implications for gamma-ray bursts *Astrophys. J.* **392** L9–13

Eichler D, Livio M, Piran T and Schramm D N 1989 Nucleosynthesis, neutrino bursts and gamma-rays from coalescing neutron stars *Nature* **340** 126–8

Filippenko A V 1988 Supernova 1987K—type II in youth, type Ib in old age *Astronom. J.* **96** 1941–8

Filippenko A V 2005 Supernovae and their massive star progenitors *Soc. Pac. Conf. Ser.* **332** 34

Fink M, Hillebrandt W and Röpke F K 2007 Double-detonation supernovae of sub-Chandrasekhar mass white dwarfs *Astron. Astrophys.* **476** 1133–43

Frail D A *et al* 2001 Beaming in gamma-ray bursts: evidence for a standard energy reservoir *Astrophys. J.* **562** L55–8

Fryer C L, Woosley S E and Hartmann D H 1999a Formation rates of black hole accretion disk gamma-ray bursts *Astrophys. J.* **526** 152–77

Fryer C L, Woosley S E, Herant M and Davies M B 1999b Merging white dwarf/black hole binaries and gamma-ray bursts *Astrophys. J.* **520** 650–60

Fynbo J P U *et al* 2006 No supernovae associated with two long-duration γ-ray bursts *Nature* **444** 1047–9

Galama T J *et al* 1998 An unusual supernova in the error box of the γ-ray burst of 25 April 1998 *Nature* **395** 670–2

Gehrels N, Ramirez-Ruiz E and Fox D B 2009 Gamma-ray bursts in the *Swift* era *Annu. Rev. Astron. Astrophys.* **47** 567–617

Giannios D 2006 Prompt emission spectra from the photosphere of a GRB *Astron. Astrophys.* **457** 763–70

Goldstein A, Connaughton V, Briggs M S and Burns E 2016 Estimating long GRB jet opening angles and rest-frame energetics *Astrophys. J.* **818** 18

Gompertz B and Fruchter A 2017 Magnetars in ultra-long gamma-ray bursts and GRB 111209A *Astrophys. J.* **839** 49

Goodman J 1986 Are gamma-ray bursts optically thick? *Astrophys. J.* **308** L47–50

Greiner J *et al* 2015 A very luminous magnetar-powered supernova associated with an ultra-long γ-ray burst *Nature* **523** 189–92

Hamuy M 2003 Observed and physical properties of core-collapse supernovae *Astrophys. J.* **582** 905–14

Hartmann D H and Woosley S E 1995 Models for classical gamma-ray bursts *Adv. Space Res.* **15** 143–52

Hjorth J *et al* 2003 A very energetic supernova associated with the γ-ray burst of 29 March 2003 *Nature* **423** 847–50

Humphreys R and Stanek K 2005 The fate of the most massive stars, *Proc. of the Conf. held 23–28 May, 2004 in Grand Teton National Park, WY* (ASP Conference Series vol 332) (San Francisco, CA: Astronomical Society of the Pacific) p.34

Ioka K and Nakamura T 2001 Peak luminosity-spectral lag relation caused by the viewing angle of the collimated gamma-ray bursts *Astrophys. J.* **554** L163–7

James P A and Anderson J P 2006 The Hα galaxy survey. III. Constraints on supernova progenitors from spatial correlations with Hα emission *Astron. Astrophys.* **453** 57–65

Jerkstrand A, Timmes F X, Magkotsios G, Sim S A, Fransson C, Spyromilio J, Müller B, Heger A, Sollerman J and Smartt S J 2015 Constraints on explosive silicon burning in core-collapse supernovae from measured Ni/Fe ratios *Astrophys. J.* **807** 110

Katz J I 1997 Yet another model of gamma-ray bursts *Astrophys. J.* **490** 633–41

Kerr R P 1963 Gravitational field of a spinning mass as an example of algebraically special metrics *Phys. Rev. Lett.* **11** 237–8

Kobayashi S, Piran T and Sari R 1997 Can internal shocks produce the variability in gamma-ray bursts? *Astrophys. J.* **490** 92

Kumar P and Piran T 2000 Some observational consequences of gamma-ray burst shock models *Astrophys. J.* **532** 286–93

Lazzati D and Begelman M C 2010 Non-thermal emission from the photospheres of gamma-ray burst outflows. I. High-frequency tails *Astrophys. J.* **725** 1137–45

Le T and Dermer C D 2007 On the redshift distribution of gamma-ray bursts in the *Swift* era *Astrophys. J.* **661** 394–415

Li W *et al* 2011 Nearby supernova rates from the Lick Observatory Supernova Search—II. The observed luminosity functions and fractions of supernovae in a complete sample *Mon. Not. R. Astron. Soc.* **412** 1441–72

Lithwick Y and Sari R 2001 Lower limits on Lorentz factors in gamma-ray bursts *Astrophys. J.* **555** 540–5

Lunnan R *et al* 2014 Hydrogen-poor superluminous supernovae and long-duration gamma-ray bursts have similar host galaxies *Astrophys. J.* **787** 138

MacFadyen A I and Woosley S E 1999 Collapsars: gamma-ray bursts and explosions in 'failed supernovae' *Astrophys. J.* **524** 262–89

MacFadyen A I, Woosley S E and Heger A 2001 Supernovae, jets, and collapsars *Astrophys. J.* **550** 410–25

Malesani D *et al* 2004 SN 2003lw and GRB 031203: a bright supernova for a faint gamma-ray burst *Astrophys. J.* **609** L5–8

Mao S and Yi I 1994 Relativistic beaming and gamma-ray bursts *Astrophys. J.* **424** L131–4

Mészáros P 2002 Theories of gamma-ray bursts *Annu. Rev. Astron. Astrophys.* **40** 137–69

Mészáros P 2006 Gamma-ray bursts *Rep. Prog. Phys.* **69** 2259–321

Mészáros P and Rees M J 1997 Optical and long-wavelength afterglow from gamma-ray bursts *Astrophys. J.* **476** 232–7

Mészáros P and Rees M J 2001 Collapsar jets, bubbles, and Fe lines *Astrophys. J.* **556** L37–40

Mészáros P and Rees M J 2015 Gamma-ray bursts *General Relativity and Gravitation: A Centennial Perspective* ed A Ashtekar *et al* (Cambridge: Cambridge University Press) pp 148–61

Metzger B D, Giannios D, Thompson T A, Bucciantini N and Quataert E 2011 The protomagnetar model for gamma-ray bursts *Mon. Not. R. Astron. Soc.* **413** 2031–56

Miralda-Escudé J 1999 Small-angle scattering of x-rays from extragalactic sources by dust in intervening galaxies *Astrophys. J.* **512** 21–4

Nakar E 2007 Short-hard gamma-ray bursts *Phys. Rep.* **442** 166–236

Nakar E and Piran T 2002 Time-scales in long gamma-ray bursts *Mon. Not. R. Astron. Soc.* **331** 40–4

Narayan R, Paczynski B and Piran T 1992 Gamma-ray bursts as the death throes of massive binary stars *Astrophys. J.* **395** L83–6

Narayan R, Piran T and Kumar P 2001 Accretion models of gamma-ray bursts *Astrophys. J.* **557** 949–57

Newman E T, Couch E, Chinnapared K, Exton A, Prakash A and Torrence R 1965 Metric of a rotating, charged mass *J. Math. Phys.* **6** 918–9

Nomoto K and Leung S-C 2018 Single degenerate models for type Ia supernovae: Progenitor's evolution and nucleosynthesis yields *Space Sci. Rev.* **214** 67

Norris J P and Bonnell J T 2006 Short gamma-ray bursts with extended emission *Astrophys. J.* **643** 266–75

Oates S R, Page M J, De Pasquale M, Schady P, Breeveld A A, Holland S T, Kuin N P M and Marshall F E 2012 A correlation between the intrinsic brightness and average decay rate of *Swift*/UVOT gamma-ray burst optical/ultraviolet light curves *Mon. Not. R. Astron. Soc.* **426** L86–90

Oates S R, Racusin J L, De Pasquale M, Page M J, Castro-Tirado A J, Gorosabel J, Smith P J, Breeveld A A and Kuin N P M 2015 Exploring the canonical behaviour of long gamma-ray bursts using an intrinsic multiwavelength afterglow correlation *Mon. Not. R. Astron. Soc.* **453** 4121–35

Ouyed R and Staff J 2013 Quark-novae in neutron star–white dwarf binaries: a model for luminous (spin-down powered) sub-Chandrasekhar-mass type Ia supernovae? *Res. Astron. Astrophys.* **13** 435–64

Paczynski B 1986 Gamma-ray bursters at cosmological distances *Astrophys. J.* **308** L43–6

Paczynski B 1990 Super-Eddington winds from neutron stars *Astrophys. J.* **363** 218–26

Paczynski B 1991 Cosmological gamma-ray bursts *Acta Astron.* **41** 257–67

Paczyński B 1998 Are gamma-ray bursts in star-forming regions? *Astrophys. J.* **494** L45–8

Page D N and Thorne K S 1974 Disk-accretion onto a black hole. Time-averaged structure of accretion disk *Astrophys. J.* **191** 499–506

Panaitescu A and Kumar P 2000 Analytic light curves of gamma-ray burst afterglows: homogeneous versus wind external media *Astrophys. J.* **543** 66–76

Panaitescu A, Spada M and Mészáros P 1999 Power density spectra of gamma-ray bursts in the internal shock model *Astrophys. J.* **522** L105–8

Panaitescu A, Mészáros P, Burrows D, Nousek J, Gehrels N, O'Brien P and Willingale R 2006 Evidence for chromatic x-ray light-curve breaks in *Swift* gamma-ray burst afterglows and their theoretical implications *Mon. Not. R. Astron. Soc.* **369** 2059–64

Pe'er A 2008 Temporal evolution of thermal emission from relativistically expanding plasma *Astrophys. J.* **682** 463–73

Pe'er A 2015 Theory of photospheric emission in gamma-ray bursts, *Thirteenth Marcel Grossmann Meeting: On Recent Developments in Theoretical and Experimental General Relativity, Astrophysics and Relativistic Field Theories* ed K Rosquist *et al* (Singapore: World Scientific) 1745–7

Piran T 1999 Gamma-ray bursts and the fireball model *Phys. Rep.* **314** 575–667

Piran T 2004 The physics of gamma-ray bursts *Rev. Mod. Phys.* **76** 1143–210

Pontzen A, Deason A, Governato F, Pettini M, Wadsley J, Quinn T, Brooks A, Bellovary J and Fynbo J P U 2010 The nature of HI absorbers in gamma-ray burst afterglows: clues from hydrodynamic simulations *Mon. Not. R. Astron. Soc.* **402** 1523–35

Rees M J and Meszaros P 1992 Relativistic fireballs—energy conversion and time-scales *Mon. Not. R. Astron. Soc.* **258** 41–3

Rees M J and Meszaros P 1994 Unsteady outflow models for cosmological gamma-ray bursts *Astrophys. J.* **430** L93–6

Rees M J and Mészáros P 1998 Refreshed shocks and afterglow longevity in gamma-ray bursts *Astrophys. J.* **496** L1–4

Rhoads J E 1999 The dynamics and light curves of beamed gamma-ray burst afterglows *Astrophys. J.* **525** 737–49

Rowlinson A, O'Brien P T and Tanvir N R 2011 The unusual x-ray emission of the short *Swift* GRB 090515: evidence for the formation of a magnetar? *AIP Conf. Ser.* **1358** 195–98

Rowlinson A, O'Brien P T, Metzger B D, Tanvir N R and Levan A J 2013 Signatures of magnetar central engines in short GRB light curves *Mon. Not. R. Astron. Soc.* **430** 1061–87

Rowlinson A, Gompertz B P, Dainotti M, O'Brien P T, Wijers R A M J and van der Horst A J 2014 Constraining properties of GRB magnetar central engines using the observed plateau luminosity and duration correlation *Mon. Not. R. Astron. Soc.* **443** 1779–87

Ruderman M and Cheng K S 1988 Evolution of a short-period gamma-ray pulsar family—Crab, Vela. COS B source, gamma-ray burst source *Astrophys. J.* **335** 306–18

Ryan G, van Eerten H, MacFadyen A and Zhang B-B 2015 Gamma-ray bursts are observed off-axis *Astrophys. J.* **799** 3

Sari R and Mészáros P 2000 Impulsive and varying injection in gamma-ray burst afterglows *Astrophys. J.* **535** L33–7

Sari R and Piran T 1997 Variability in gamma-ray bursts: a clue *Astrophys. J.* **485** 270–3

Sari R, Piran T and Halpern J P 1999 Jets in gamma-ray bursts *Astrophys. J.* **519** L17–20

Shao L and Dai Z G 2007 Behavior of x-ray dust scattering and implications for x-ray afterglows of gamma-ray bursts *Astrophys. J.* **660** 1319–25

Shao L, Dai Z G and Mirabal N 2008 Echo emission from dust scattering and x-ray afterglows of gamma-ray bursts *Astrophys. J.* **675** 507–18

Shemi A and Piran T 1990 The appearance of cosmic fireballs *Astrophys. J.* **365** L55–8

Shen R-F, Willingale R, Kumar P, O'Brien P T and Evans P A 2009 The dust scattering model cannot explain the shallow x-ray decay in GRB afterglows *Mon. Not. R. Astron. Soc.* **393** 598–606

Smartt S J, Vreeswijk P M, Ramirez-Ruiz E, Gilmore G F, Meikle W P S, Ferguson A M N and Knapen J H 2002 On the progenitor of the type Ic supernova 2002ap *Astrophys. J.* **572** L147–51

Stanek K Z, Garnavich P M, Kaluzny J, Pych W and Thompson I 1999 BVRI observations of the optical afterglow of GRB 990510 *Astrophys. J.* **522** L39–42

Thorne K S 1974 Disk-accretion onto a black hole. II. Evolution of the hole *Astrophys. J.* **191** 507–20

Thorne K S, Price R H and MacDonald D A 1986 *Black Holes: The Membrane Paradigm* (New Haven, CT: Yale University Press)

Toy V L *et al* 2016 Optical and near-infrared observations of SN 2013dx associated with GRB 130702A *Astrophys. J.* **818** 79

Troja E *et al* 2016 An achromatic break in the afterglow of the short GRB 140903A: evidence for a narrow jet *Astrophys. J.* **827** 102

Urata Y *et al* 2007 Testing the external-shock model of gamma-ray bursts using the late-time simultaneous optical and x-ray afterglows *Astrophys. J.* **668** L95–8

Usov V V 1992 Millisecond pulsars with extremely strong magnetic fields as a cosmological source of gamma-ray bursts *Nature* **357** 472–4

Van Dyk S D, Li W and Filippenko A V 2003 A search for core-collapse supernova progenitors in *Hubble Space Telescope* Images *Publ. Astron. Soc. Pac.* **115** 1–20

Vietri M and Stella L 1998 A gamma-ray burst model with small baryon contamination *Astrophys. J.* **507** L45–8

Woosley S E 1993 Gamma-ray bursts from stellar mass accretion disks around black holes *Astrophys. J.* **405** 273–7

Woosley S E and Bloom J S 2006 The supernova gamma-ray burst connection *Annu. Rev. Astron. Astrophys.* **44** 507–56

Wu X-F, Hou S-J and Lei W-H 2013 Giant x-ray bump in GRB 121027A: evidence for fall-back disk accretion *Astrophys. J.* **767** L36

Xin L-P *et al* 2016 Multi-wavelength observations of GRB 111228A and implications for the fireball and its environment *Astrophys. J.* **817** 152

Yamazaki R, Yonetoku D and Nakamura T 2003 An off-axis jet model for GRB 980425 and low-energy gamma-ray bursts *Astrophys. J.* **594** L79–82

Yu Y-W, Zhu J-P, Li S-Z, Lü H-J and Zou Y-C 2017 A statistical study of superluminous supernovae using the magnetar engine model and implications for their connection with gamma-ray bursts and hypernovae *Astrophys. J.* **840** 12

Zhang B 2007 Gamma-ray bursts in the *Swift* era *Chinese J. Astron. Astrophys.* **7** 1–50

Zhang B and Mészáros P 2001 Gamma-ray burst afterglow with continuous energy injection: signature of a highly magnetized millisecond pulsar *Astrophys. J.* **552** L35–8

Zhang B and Mészáros P 2002 An analysis of gamma-ray burst spectral break models *Astrophys. J.* **581** 1236–47

Zhang W, Woosley S E and MacFadyen A I 2003 Relativistic jets in collapsars *Astrophys. J.* **586** 356–71

Zhang Z, Xie G Z, Deng J G and Jin W 2006 Revisiting the characteristics of the spectral lags in short gamma-ray bursts *Mon. Not. R. Astron. Soc.* **373** 729–32

Chapter 3

GRB correlations between prompt parameters

3.1 Why are standard candles and sirens important for cosmology?

One of the main problems in astronomy is the computation of distances and there are a lot of methods, applied at different distance ranges. Parallax measurements are used for computing the distances of nearby stars, the periodicity of Cepheids is used for deriving the distances of nearby galaxies, while for more distant objects the method of SNe Ia, which are the most popular standard candles, is used. Standard candles are astronomical objects for which the luminosity is known or can be computed through other well-established correlations. Consequently, the luminosity distance can be derived through the luminosity of the objects or through these correlations. Standard sirens are astrophysical objects for which the luminosity distance is computed from the waveform created by a binary system. Thus, GW sources are standard distance indicators. For GWs the main issue is to calculate the redshift of the object through an electromagnetic (EM) counterpart, such as EM emission from a merger or through the hosting galaxy. Here a comparison between standard candles and standard sirens is made, pinpointing the advantages and drawbacks.

Using EM waves:
- Measuring redshift is easy through the comparison of EM spectra.
- Measuring distance is difficult, because objects of known luminosity such as SNe Ia are needed.

Using GWs:
- Measuring distance is simple, because it can be derived directly from the waveform (standard sirens).
- Measuring redshift is difficult, because the problem of mass degeneracy in the waveform has to be addressed. The main aim is to identify an EM counterpart, such as optical, radio, X-ray or gamma-ray emission, together with its precise distance from the GW to distinguish the object or the host galaxy.

GWs were not observed until 2016 (Abbott *et al* 2016), because they do not interact through electromagnetic forces, and instruments with sufficient sensitivity were lacking. Nevertheless, Schutz (1986) suggested that with the instruments in preparation, the GW signal coming from the coalescence of an NS binary system had the potential to constrain the value of the present-day Hubble constant, H_0. claimed that the observations of GWs from binary systems through future GW detectors, in particular the LISA mission, will be able to determine the beginning of the star formation epoch and, together with telescopes providing electromagnetic observations, to measure the expansion of the galaxies with high precision.

In addition, Holz and Hughes (2005) suggested that the GWs coming from a coalescent supermassive binary BH system and detected by the future LISA mission will be able to compute luminosity distances with 1%–10% accuracy. In particular, with an additional EM counterpart, those observations will also allow the measurement of the redshift; thus coalescent binary BH systems could be conceived as standard candles.

Recently, Abbott *et al* (2016) claimed the detection of the first GW signal from a binary BH system by the LIGO and VIRGO collaborations, without an EM counterpart, however. Nevertheless, this opened a new branch of study in astronomy. As a further step, Abbott *et al* (2017a) presented the detection of the first GW signal from an NS binary system by the LIGO and VIRGO collaborations with a following optical counterpart detected by other space missions. Later, from the analysis of Abbott *et al* (2017b), the value of H_0 was found using only the GW signal associated with one GRB, and it was found to be in agreement with the previous values present in the literature. Indeed, they found $H_0 = 70.0^{+12.0}_{-8.0}$ km s^{-1} Mpc^{-1}, in agreement with the estimate (67.74 ± 0.46 km s^{-1} Mpc^{-1}) from Planck (Ade *et al* 2016) and the value (73.24 ± 1.74 km s^{-1} Mpc^{-1}) from SHoES observations of SNe Ia (Riess *et al* 2016). This result will allow the measurement of the luminosity distance up to cosmological scales, using just the GW signal. GRBs are good candidates for standard candles because they would allow for an extension of the Hubble diagram (HD) an order of magnitude further than SNe Ia. In fact, GRBs are detected up to redshift $z = 9.4$ (Cucchiara *et al* 2011), while SNe Ia are detected up to $z = 2.26$ (Rodney *et al* 2015). This is very helpful for analysing the dark energy (DE). SNe are well known standard candles and their use to measure the expansion of the Universe (Riess *et al* 1998, Perlmutter *et al* 1998) was worthy of the Nobel Prize in 2011. The estimation through SNe Ia of H_0 is even more precise if associated with cosmic microwave background radiation (CMBR) and baryon acoustic oscillation (BAO) estimations (Weinberg *et al* 2013). For this reason, evaluations from GRBs would also validate or limit the parameter H_0. The isotropic energy of GRBs extends over eight orders of magnitude (Lin *et al* 2015), and these phenomena can trace the cosmic SFR (Totani 1997, Porciani and Madau 2011, Bromm and Loeb 2006, Kistler *et al* 2009, de Souza *et al* 2011) and report on the physics of the intergalactic medium (Barkana and Loeb 2004, Ioka and Mészáros 2005, Inoue *et al* 2007). Therefore, GRB correlations are potentially extremely powerful tools to explain the physics of GRBs, define a good distance indicator and examine the Universe at high redshift (Salvaterra 2015).

3.2 Notations, nomenclature and abbreviations

Here, a summary of the nomenclature employed in this book is presented. This is mostly taken from the nomenclature used in Dainotti *et al* (2018) and Dainotti and Del Vecchio (2017). L, F, E, S and T represent the luminosity, the flux, the energy, the fluence and the timescale, respectively. The first subscript indicates the wavelength in which they are computed, while the second shows the different times or part of the light curve. L indicates the luminosity, and in particular L_{peak} and L_{iso} represent the peak luminosity and the total isotropic luminosity in a given energy band. L_{peak} is computed in this way:

$$L_{peak} = 4\pi D_L(z, \Omega_M, \Omega_\Lambda)^2 F_{peak}, \tag{3.1}$$

with $D_L(z, \Omega_M, \Omega_\Lambda)$ the luminosity distance given by

$$D_L(z, \Omega_M, \Omega_\Lambda) = \frac{c(1+z)}{H_0} \int_0^z \frac{dz'}{\sqrt{\Omega_M(1+z')^3 + \Omega_\Lambda}}, \tag{3.2}$$

with Ω_M and Ω_Λ the matter and dark energy density parameters, and z the redshift. Similarly, L_{iso} is obtained:

$$L_{iso} = 4\pi D_L(z, \Omega_M, \Omega_\Lambda)^2 F_{tot}. \tag{3.3}$$

Furthermore, α, β and ν, indicate the temporal and spectral decay indices and the frequencies.

Here the nomenclature is shown in detail:

- T_{90} is the time in which 90% of the GRB's fluence, starting from 5%, is radiated (Kouveliotou *et al* 1993).
- T_{45} is the time in which 45% of the total counts are detected above the background (Reichart *et al* 2001).
- T_{break} represents the break time in the afterglow light curve (Sari *et al* 1999, Willingale *et al* 2010).
- T_{peak} is the time at which a given pulse (Fishman *et al* 1994, Stern and Svensson 1996, Ryde and Svensson 2002) in the prompt light curve peaks.
- τ_{lag} and τ_{RT} are the time differences in the arrival to the observer of the high-energy (100–300 keV) photons and low-energy (25–50 keV) photons, and the shortest time over which the light curve rises by 50% of the peak flux of the pulse.
- $T_{X,a}$ and $T_{O,a}$ are the time in the X-ray wavelength at the end of the plateau and the equivalent in the optical wavelength, respectively. $F_{X,a}$ and $F_{O,a}$ represent their respective fluxes, while $L_{X,a}$ and $L_{O,a}$ denote their respective luminosities. The energy of the plateau can be approximated by $E_{X,plateau} = (L_{X,a} \times T^*_{X,a})$.

- $T_{O,\mathrm{peak}}$ and $T_{X,f}$ represent the peak time in the optical band and the time since ejection of the pulse (T_{ej}). $L_{O,\mathrm{peak}}$ and $L_{X,f}$ indicate their respective luminosities. $F_{O,\mathrm{peak}}$ is the flux at time $T_{O,\mathrm{peak}}$.
- $T_{X,\mathrm{peak}}$ defines the peak time in the X-ray range and $F_{X,\mathrm{peak}}$ and $L_{X,\mathrm{peak}}$ indicate its flux and luminosity, respectively.
- $T_{X,p}$ and $T_{X,t}$ display the time at the end of the prompt emission in the W07 model and the time at which the flat and the step decay trends of the light curves join, respectively.
- E_{peak}, E_{iso}, E_{γ} and E_{prompt} are the spectral peak energy (Mallozzi *et al* 1995), the total isotropic energy emitted during the whole burst (e.g., Amati *et al* 2002), the total energy corrected for the beaming factor and the isotropic energy emitted in the prompt phase, respectively.
- F_{peak}, F_{tot} are the peak and total fluxes, respectively (Lee and Petrosian 1996).
- $L_{X,p}$ is the luminosity at time $T_{X,p}$ in the X-ray band.
- $L_{X,200\mathrm{s}}$, $L_{X,10}$, $L_{X,11}$, $L_{X,12}$, $L_{X,1\mathrm{d}}$ and $L_{O,200\mathrm{s}}$, $L_{O,10}$, $L_{O,11}$, $L_{O,12}$, $L_{O,1\mathrm{d}}$ represent the X-ray and optical luminosities at 200 s, at 10, 11, 12 h and at 1 day, respectively; $L_{O,100\mathrm{s}}$, $L_{O,1000\mathrm{s}}$, $L_{O,10\,000\mathrm{s}}$ and $L_{O,7}$ denote the optical luminosity at 100 s, 1000 s, 10 000 s and 7 h; $L_{\gamma,\mathrm{iso}}$ and $L_L(\nu, T_{X,a})$ are the isotropic prompt emission mean luminosity and the optical or X-ray luminosity of the late prompt emission at time $T_{X,a}$.
- $F_{X,11}$, $F_{X,1\mathrm{d}}$ and $F_{O,11}$, $F_{O,1\mathrm{d}}$ indicate the X-ray and optical fluxes at 11 h and at 1 day, respectively.
- $F_{\gamma,\mathrm{prompt}}$, $F_{X,\mathrm{afterglow}}$ represent the gamma-ray flux in the prompt and the X-ray flux in the afterglow phase, respectively. $E_{\gamma,\mathrm{prompt}}$ and $E_{X,\mathrm{afterglow}}$ indicate their respective isotropic energies, while $L_{\gamma,\mathrm{prompt}}$ and $L_{X,\mathrm{afterglow}}$ indicate their respective luminosities. $S_{\gamma,\mathrm{prompt}}$ is the prompt fluence in the gamma band correspondent to the rest-frame isotropic prompt energy $E_{\gamma,\mathrm{prompt}}$.
- $E_{O,\mathrm{afterglow}}$, $E_{\gamma,\mathrm{iso}}$ and $E_{X,f}$ represent the optical isotropic energy in the afterglow phase, the total gamma isotropic energy and the break energy of the pulse.
- $E_{k,\mathrm{aft}}$ and $E_{\gamma,\mathrm{peak}}$ represent the isotropic kinetic afterglow energy in the X-ray band and the prompt peak energy in the νF_{ν} spectrum.
- $\alpha_{X,a}$, $\alpha_{O,>200\mathrm{s}}$, $\alpha_{X,>200\mathrm{s}}$, $\alpha_{\nu,\mathrm{fl}}$ and $\alpha_{\nu,\mathrm{st}}$ indicate the X-ray temporal decay index in the afterglow phase, in the optical band after 200 s, in the X-ray band after 200 s, and the optical or X-ray flat and steep temporal decay indices, respectively.
- $\beta_{X,a}$, $\beta_{OX,a}$ and $\beta_{O,>200\mathrm{s}}$ represent the spectral index of the plateau emission in the X-ray band, the optical-to-X-ray spectral index for the end time of the plateau and the optical spectral index after 200 s.
- ν_X, ν_O, ν_c, ν_m designate the X-ray and optical frequencies, and the cooling and the peak frequencies of the synchrotron radiation.
- S_{obs}, S_{tot} represent the observed fluence in the range 50–300 keV, and the total fluence in the 20 keV–1.5 MeV energy band.

- V indicates the variability of the GRB's light curve. It is calculated from the difference between the observed light curve and its smoothed version, squaring and adding these squared differences over time intervals, and normalizing the result of this sum (Reichart et al 2001). Different smoothing filters may be applied (see also Li and Paczyński 2006). V_f indicates the variability for a specific fraction of the smoothing timescale for the light curve.

The majority of the parameters mentioned above are presented in the observer frame, except for E_{iso}, E_{prompt}, L_{peak} and L_{iso}. The rest-frame quantities are indicated with the superscript '*', such as $T^*_{X,a} = T_{X,a}/(1 + z)$ and $E^*_{peak} = E_{peak}(1 + z)$.

The Pearson correlation coefficient (Bevington and Robinson 2003) is indicated by r, the Spearman correlation coefficient[1] (Spearman 1904) by ρ, the Kendall correlation coefficient (Kendall 1938) by τ,[2] and the p-value (the probability that a correlation is due to chance) is represented by P. Given that almost all the correlations described here are represented by power-laws, the slope is that of the equivalent log–log correlation.

3.3 The GRB correlations between prompt parameters

Here physical phenomenological correlations between important observables in GRBs are described. Each section briefly summarizes the discovery of the correlations, their discussion in the literature and their physical interpretation.

3.3.1 The L_{peak}–τ_{lag} correlation

Liang and Kargatis (1996) discovered, employing 34 bright GRBs detected by BATSE, that E_{peak} depends linearly on the instantaneous luminosity. Quantitatively,

$$\frac{L_{peak}}{N} = -\frac{dE_{peak}}{dt},$$ (3.4)

where N is a normalization constant denoting the pulse luminosity.

The L_{peak}–τ_{lag} anti-correlation was discovered by Norris et al (2000) with a sample of six GRBs with firm redshift (see the left panel of figure 3.1)

$$\log L_{peak} = 55.11 - 1.14 \log \tau^*_{lag},$$ (3.5)

with L_{peak}, in units of 10^{53} erg s^{-1}, calculated between 50 and 300 keV, and τ^*_{lag} in seconds. A correlation in agreement with the previous one was found by Schaefer et al (2001) with 112 BATSE GRBs :

---

[1] This coefficient measures the level of correlation between two quantities using a monotonic function. Its value ranges between −1 and +1.
[2] The Kendall coefficient τ calculates non-parametrically the rank correlation between two variables.

$$\log L_{\text{peak}} = 52.46 - (1.14 \pm 0.20) \log \tau_{\text{lag}}. \tag{3.6}$$

This correlation agrees well with the outcome found by Norris *et al* (2000). In this work, L_{peak} is instead in units of 10^{51} erg s^{-1}. This correlation has been an object of discussion and investigation (e.g. Salmonson 2000, Daigne and Mochkovitch 2003, Zhang *et al* 2006).

Schaefer (2004) suggested that the L_{peak}–τ_{lag} correlation is derived by the Liang and Kargatis (1996) correlation from equation (3.4). This correlation was useful (Schaefer *et al* 2001) to infer pseudo-redshifts. Kocevski and Liang (2003) varied the predicted z until the D_L computed through the estimated redshift was in agreement with the measured D_L with a precision for the convergence of these two values of 10^{-3}. This was performed by always fixing a flat ΛCDM model. Additionally, Kocevski and Liang (2003) demonstrated that the GRB's spectral evolution is related to L_{peak}. Instead, Hakkila *et al* (2008), finding a different slope (-0.62 ± 0.04), concluded that the L_{peak}–τ_{lag} correlation is a characteristic of the pulse, not a feature of a burst. Tsutsui *et al* (2008), using pseudo-redshifts estimated using the Yonetoku correlation (see section 3.3.6), found that the L_{peak}–τ_{lag} correlation has a small correlation coefficient $\rho = 0.38$ (see the right panel of figure 3.1). Taking into account the dependence of the luminosity on both redshift and τ_{lag}, they obtained

$$\log L_{\text{peak}} = 50.88 + 2.53 \log(1 + z) - 0.282 \log \tau_{\text{lag}}, \tag{3.7}$$

with L_{peak} in units of 10^{50} erg s^{-1}, τ_{lag} in seconds, $\rho = 0.77$ and $P = 7.9 \times 10^{-75}$. This outcome suggests that an eventual L_{peak}–τ_{lag} correlation recovered using *Swift* data should show dependence on the redshift.

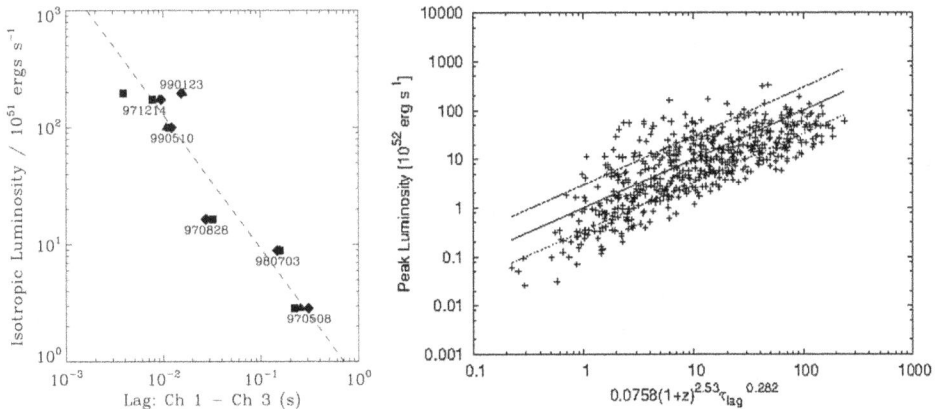

Figure 3.1. Left panel: the L_{peak}–τ^*_{lag} correlation adopting six GRBs with measured redshift. The power-law fit employing channel 1 (25–50 keV) and channel 3 (100–300 keV) of the BATSE instrument is displayed by the dashed line. (Reproduced with permission from Norris *et al* (2000). Copyright 2000 The American Astronomical Society.) Right panel: the $\log L_{\text{peak}} \sim 2.53 \log(1 + z) - 0.282 \log \tau_{\text{lag}}$ distribution with correlation coefficient $\rho = 0.77$, $P = 7.9 \times 10^{-75}$. The best-fit line is represented by the solid line. (Reproduced with permission from Tsutsui *et al* (2008). Copyright 2008 AIP Publishing.)

Following this track, Sultana *et al* (2012) presented a correlation between the z- and k-corrected τ_{lag} (computed between 50 and 100 keV and between 100 and 200 keV) and L_{peak} for 12 *Swift* LGRBs. The correction for the time dilatation effect together with the k-correction in the energy bands (Gehrels *et al* 2006) makes τ^*_{lag} equal to $(1 + z)^{-0.67}\tau_{\text{lag}}$. Sultana *et al* (2012) showed that the L_{peak}–τ^*_{lag} correlation can be the extrapolation of the LT correlation within the prompt emission. The slope of the L_{peak}–τ^*_{lag} correlation is compatible within 1σ with the slope of the LT relation (Dainotti *et al* 2008, 2010, Dainotti *et al* 2011a, 2013a).

Sultana *et al* (2012) found[3]:

$$\log L_{\text{peak}} - (54.87 \pm 0.29) - (1.19 \pm 0.17) \log\left[(1 + z)^{-0.67}\tau_{\text{lag}}\right], \tag{3.8}$$

and

$$\log L_{X,a} = (51.57 \pm 0.10) - (1.10 \pm 0.03) \log T^*_{X,a}, \tag{3.9}$$

with τ_{lag} in ms, T^*_a in seconds and L in erg s^{-1}. These two correlations both have significant correlation coefficients ($\rho = -0.65$ for the L_{peak}–τ_{lag} correlation and $\rho = -0.88$ for the LT correlation) and have analogous values of their slopes (-1.19 ± 0.17 for the L_{peak}–τ_{lag} and -1.10 ± 0.03 for the LT relation). Although τ_{lag} and $T^*_{X,a}$ represent different GRB time variables, it is clear that the L_{peak}–τ_{lag} correlation extrapolates into the LT one for timescales $\tau_{\text{lag}} \simeq T^*_{X,a}$.

Ukwatta *et al* (2010) confirmed the existence of a correlation between L^*_{peak} and both the time dilated and k-corrected τ_{lag} for 31 *Swift* GRBs. From their analysis they found $r = -0.68$, $P = 7 \times 10^{-2}$, and a slope equal to -1.4 ± 0.3, although with a large scatter. In addition, Ukwatta *et al* (2012), with a sample of 43 *Swift* GRBs with measured redshift, found $r = -0.82$, $P = 5.5 \times 10^{-5}$ and a slope of -1.2 ± 0.2, compatible with earlier findings.

Finally, Margutti *et al* (2010) concluded that the X-ray flares follow the same L_{peak}–τ^*_{lag} correlation (between 0.3 and 10 keV in the rest-frame) of the GRBs, and claimed that the emission process of both X-ray flares and GRBs is analogous.

Physical interpretation of the L_{peak}–τ_{lag} correlation
The main hypothesis of Norris *et al* (2000) is that the energy formation process is much more important than dissipation in the pulse dynamics. Investigating several pulses in the very luminous LGRB detected by BATSE, it was found that the rise-to-decay ratio is $\leqslant 1$ for pulses with a well-defined shape. Moreover, when the ratio becomes lower the pulse appears broader and dimmer. Another interpretation of the L_{peak}–τ_{lag} correlation is the one given by Salmonson (2000) for which this correlation is due to a kinematic effect. In this framework, an emitting region with constant luminosity is the origin of the GRB's emission. Additionally, in this scenario the observed luminosity should be proportional to Γ, while the apparent luminosity τ_{lag}

---

[3] In Sultana *et al* (2012) the peak isotropic luminosity is represented by L_{iso}.

is proportional to $1/\Gamma$. Nevertheless, Schaefer (2004) noted that several drawbacks are associated with this scenario:

- The variation in the velocity along the line of sight is due to the velocity of the jet expansion together with the cosmological expansion.
- The discrepancies in luminosity and τ_{lag} among GRBs are caused by the different velocities of the emitting regions. This gives consistent results with the L_{peak}–τ_{lag} correlation. However, such an explanation demands a constant comoving luminosity for all GRBs, which represents a severe requirement.
- Γ and luminosity should vary consistently. This is not the case, because the observed L_{peak} spans more than three orders of magnitude (e.g. Schaefer *et al* 2001), while Γ varies by less than one order of magnitude (Panaitescu and Kumar 2002).
- The observed luminosity should depend linearly on the parameter Γ of the jet. However, this requirement is not fulfilled. In fact, a number of corrections that produce a nonlinear dependence should be accounted for. For example, the forward jet motion adds a quadratic dependence (Fenimore *et al* 1996).

Another interpretation for the L_{peak}–τ_{lag} correlation was suggested by Ioka and Nakamura (2001). They suggested that the L_{peak} depends on the viewing angle, pointing out that a jet with a tiny viewing angle leads to a high-luminosity peak L_{peak} in GRBs with a short τ_{lag}. The viewing angle also affects other correlations (see the L_{peak}–V correlation presented in section 3.3.2). Additionally, XRFs may be considered as GRBs observed from large viewing angles with large τ_{lag} and negligible variability. However, under the assumption of particular jet angle distributions, that the rate of GRBs is the same as the SFR and accounting for selection effects, Lü *et al* (2012) found an anti-correlation between θ_{jet} and the redshift for 77 GRBs. This correlation reads as follows:

$$\log \theta_{jet} = (-0.90 \pm 0.09) - (0.94 \pm 0.19) \log(1 + z), \qquad (3.10)$$

with $\rho = 0.55$ and $P < 10^{-4}$.

Through simulations, they pointed out that the instrumental selection effects play a major role in the observed θ_{jet}–z as well as in the τ_{lag}–z correlations (Yi *et al* 2008). These biases are also relevant in the dependence on the redshift of the shallow decays in the X-ray afterglow phases (Stratta *et al* 2009). Recently, Ryan *et al* (2015) explored 226 jet opening angles of *Swift*/XRT GRB light curves with known redshift. As a result, the majority of the GRBs were observed off-axis, giving an important role to the viewing angle for the behaviour of the afterglow light curves. Regarding its physical interpretation, Zhang *et al* (2009) claimed a connection between the L_{peak}–τ_{lag} correlation and the L_{peak}–V one (see section 3.3.2), giving an explanation for the latter correlation through the internal shock scenario (see section 3.3.2). Later, Uhm and Zhang (2016) employed the synchrotron radiation mechanism to describe the values of τ_{lag} and this explanation is in agreement with observations. An additional possibility to interpret the origin of the L_{peak}–τ_{lag}

correlation is provided by Sultana *et al* (2012). They interpreted the L_{peak}–τ_{lag} correlation using the following parameter:

$$D = \frac{1}{\Gamma(1 - \beta_0 \cos \theta)(1 + z)},$$ (3.11)

where D is the Doppler factor of a jet at a viewing angle θ and with velocity $\beta_0 \equiv v/c$ at redshift z. D is also related to the rest-frame timescale τ and the observed time t as follows:

$$t = \frac{\tau}{D}.$$ (3.12)

Thus, assuming a timescale $\Delta\tau$ in the GRB rest-frame, equation (3.12) will become $\Delta t = \Delta\tau/D$ in the observer frame. Then, L_{peak}, assuming a power-law spectrum, is given by

$$L_{peak} \propto D^{\alpha},$$ (3.13)

with $\alpha \approx 1$. This result reproduced, together with equation (3.12), the L_{peak}–τ_{lag} correlation. In conclusion, a similar description for the L_{peak}–τ_{lag} and LT correlations is suggested by Sultana *et al* (2012) given the similar correlation coefficients and best-fit slopes of the two relations, as already discussed in section 3.3.1.

3.3.2 The L_{peak}–V correlation

Fenimore and Ramirez-Ruiz (2000) found the following correlation between L_{peak} and V:

$$\log L_{peak} = 56.49 + 3.35 \log V,$$ (3.14)

L_{peak}, in erg s^{-1}, is computed in the rest-frame between 50 and 300 keV. After calibrating the correlation, a sample of 220 GRBs detected by BATSE were used to compute luminosities, distances and the GRB formation rate (GFR). Nevertheless, the authors state that further analysis for the validation of this L_{peak}–V correlation was needed.

Employing a sample of 20 GRBs detected by *CGRO*/BATSE, KONUS/*Wind* and *Ulysses*/GRB, Reichart *et al* (2001) discovered the following correlation:

$$\log L_{peak} \sim \left(3.3^{+1.1}_{-0.9}\right) \log V,$$ (3.15)

with $\rho = 0.8$ and $P = 1.4 \times 10^{-4}$ (see the left panel of figure 3.2); L_{peak} was calculated between 50 and 300 keV in the observer frame. The $\log L_{peak}$–$\log V_f$ correlation can be represented in this way:

$$\log V_f(L) = \log \bar{V}_f + b + m(\log L_{peak} - \log \bar{L}_{peak}),$$ (3.16)

where $b = 0.013^{+0.075}_{-0.092}$, $m = 0.302^{+0.112}_{-0.075}$, and $\bar{V}_f$ and $\bar{L}_{peak}$ are the median values of V_f and L_{peak}.

An update of the $\log L_{\text{peak}}$–$\log V$ correlation was described in Guidorzi *et al* (2005). Employing 32 GRBs detected by *BeppoSAX*, *CGRO*/BATSE, *HETE-2* and KONUS, they found the following correlation (see the right panel of figure 3.2):

$$\log L_{\text{peak}} = 3.36^{+0.89}_{-0.43} + 1.30^{+0.84}_{-0.44} \log V, \tag{3.17}$$

with $\rho = 0.625$ and $P = 10^{-4}$, and L_{peak} in units of 10^{50} erg s^{-1}.

The last result was criticized by Reichart and Nysewander (2005), who pointed out that in Guidorzi *et al* (2005) the variance of the sample in their fit was not taken into account. Reichart and Nysewander (2005), with an updated dataset, confirmed the results obtained in Reichart *et al* (2001), finding a correlation with slope $3.4^{+0.9}_{-0.6}$ and variance $\sigma_V = 0.2 \pm 0.04$.

Later, Guidorzi *et al* (2006) analysed the $L_{\text{peak}} - V$ correlation by applying the statistical fitting methods described in Reichart *et al* (2001) and D'Agostini (2005). The former method recovered a L_{peak}–V correlation compatible with earlier works:

$$\log L_{\text{peak}} \sim 3.5^{+0.6}_{-0.4} \log V. \tag{3.18}$$

In contrast, the latter method found a correlation slope lower that from Reichart *et al* (2001):

$$\log L_{\text{peak}} \sim 0.88^{+0.12}_{-0.13} \log V. \tag{3.19}$$

The latter slope is in agreement with the outcomes of Guidorzi *et al* (2005), but incompatible with those of Reichart and Nysewander (2005).

Rizzuto *et al* (2007), using 36 *Swift* LGRBs with known redshifts and $L_{\text{peak}} > 5 \times 10^{50}$ erg s^{-1} between 100 and 1000 keV, investigated the L_{peak}–V

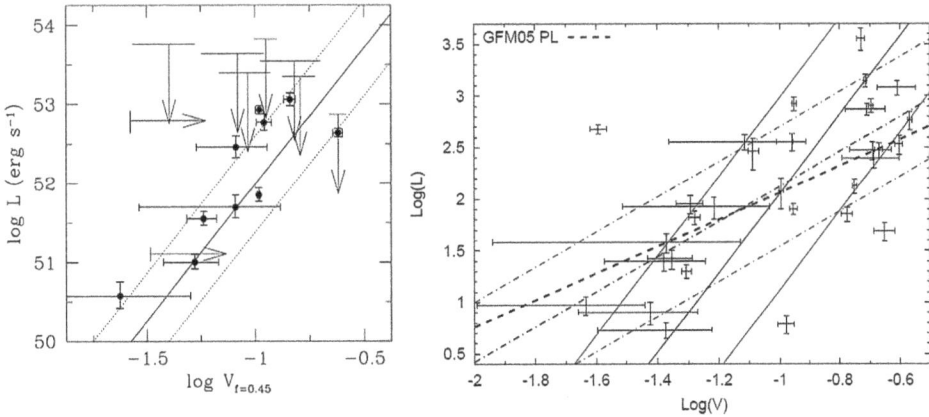

Figure 3.2. Left panel: the L_{peak}–$V_{f=0.45}$ correlation with the solid and dotted lines the best-fit line and 1σ deviation, respectively. (Reproduced with permission from Reichart *et al* (2001). Copyright 2001 The American Astronomical Society.) Right panel: the $\log L_{\text{peak}}$–$\log V$ correlation for a dataset of 32 GRBs. The solid lines are obtained through the Reichart *et al* (2001) method, dashed-dotted lines through the D'Agostini (2005) method and the dashed lines are from Guidorzi *et al* (2005). (Reproduced from Guidorzi *et al* (2006). By permission of Oxford University Press on behalf of the Royal Astronomical Society.)

correlation. Rizzuto *et al* (2007) assumed two definitions for the variability V: the first by Reichart *et al* (2001), V_R, and the latter by Li and Paczyński (2006), V_{LP}, which selects only high-frequency variability. In this way, Rizzuto *et al* (2007) obtained:

$$\log L_{\text{peak}} \sim (2.3 \pm 0.17) \log V_{LP}, \tag{3.20}$$

with $\rho = 0.758$ and $P = 0.011$, and

$$\log L_{\text{peak}} \sim (1.7 \pm 0.4) \log V_R, \tag{3.21}$$

with $\sigma_{\log L} = 0.58^{+0.15}_{-0.12}$, $\rho = 0.115$ and $P = 0.506$. From their work, six dim GRBs (GRB 050223, GRB 050416A, GRB 050803, GRB 051016B, GRB 060614 and GRB 060729) were classified as outliers of the correlation due to their high values of V_R. For this reason, the correlation appears to be valid for luminous GRBs only. In conclusion, the variance in this correlation is more significant than in other ones. Thus, this outcome shows that this correlation is less robust. Nevertheless, it is worth examining the physical interpretation of this correlation for additional studies.

Physical interpretation of the L_{peak}–V *correlation*
In the framework of the internal and external shock model (Piran 2004, Mészáros 2006, Fenimore and Ramirez-Ruiz 2000) it was pointed out that the interpretation of the L_{peak}–V correlation is not clear. Indeed, larger initial Γ values should result in more efficient collisions. After modifying several variables, i.e. Γ, the initial shells' mass and the external medium density, it is not possible to reproduce the observed variability V. Therefore, the central engine seems to be crucial in the description of this correlation. Salmonson and Galama (2002) investigated the correlation assuming that the GRB variability is due to a variation in the value of the opening angles of the jets. Small values of θ_{jet} produce faster outflows. They found out that under these assumptions it is possible to explain high luminosities, high variability values, short pulse lags and a jet break at early time for GRBs observed on-axis. Instead, low luminosities, flatter light curves, broader pulse lags and jet breaks at later times suggest larger viewing angles.

Guidorzi *et al* (2006) suggested that a stronger dependence of Γ on θ_{jet} is needed to interpret this correlation within the jet-emission model. However, given that V and L_{peak} are proportional to Γ and fast rise times and short pulse durations lead to high variability, Schaefer (2007) attributed the origin of the L_{peak}–V correlation to relativistically shocked jets.

3.3.3 The L_{iso}–τ_{RT} correlation and its physical interpretation

Schaefer (2003b) proposed the correlation between τ_{RT} and L_{iso}:

$$L_{\text{iso}} \propto \tau_{RT}^{-N/2}, \tag{3.22}$$

with the exponent $N \simeq 3$ (see also Schaefer 2003b, 2007). From this correlation it follows that high luminosities are produced by fast rises and low luminosities from slow rises. Regarding the physics of the shocked jet, τ_{RT} represents the largest delay between the time of the arrival of photons from the centre of the visible region and their arrival time from its edge.

This delay generating the $\tau_{RT} \propto \Gamma^{-2}$ correlation is due to the angular opening of the emitted jet. Panaitescu and Kumar (2002) claimed the existence of a minimum radius similar for all the GRBs under which the emission from the material is not efficient anymore.

Using the common dependence on Γ of τ_{RT} and L_{iso}, Schaefer (2007) confirmed that $\log L_{iso}$ is $\sim -N/2 \log \tau_{RT}$. With a sample of 69 BATSE and *Swift* GRBs, the correlation reads as follows:

$$\log L_{iso} = 53.54 - 1.21 \log \tau_{RT}^{*}, \qquad (3.23)$$

with L_{iso} in erg s^{-1} and τ_{RT}^{*} in seconds. The 1σ errors are represented in the left panel of figure 3.3. The error in $\log L_{iso}$ is

$$\sigma_{\log L_{iso}}^{2} = \sigma_{a}^{2} + \left[\sigma_{b} \log \frac{\tau_{RT}^{*}}{0.1 \text{ s}} \right]^{2} + \left(\frac{0.43 b \sigma_{RT}}{\tau_{RT}} \right)^{2} + \sigma_{RT,sys}^{2}, \qquad (3.24)$$

where Schaefer (2007) took into account an additional scatter, σ_{sys}. When $\sigma_{RT,sys} = 0.47$, the χ^{2} of the best-fit is equal to 1. With a sample of 107 GRBs detected by BATSE, *HETE-2*, KONUS and *Swift*, with measured spectroscopic redshift (see the right panel of figure 3.3), Xiao and Schaefer (2009) pointed out, accounting also for the Poisson noise, the following correlation:

$$\log L_{iso} = 53.84 - 1.70 \log \tau_{RT}^{*}, \qquad (3.25)$$

with the same units as in equation (3.23).

This analysis showed that the smoothing of the light curve to calculate the τ_{RT} is relevant because it can lead to underestimation/overestimation of this parameter. This problem can be very significant for low-luminosity bursts. In conclusion, Schaefer (2007) provided the physical interpretation of this correlation. Indeed, in

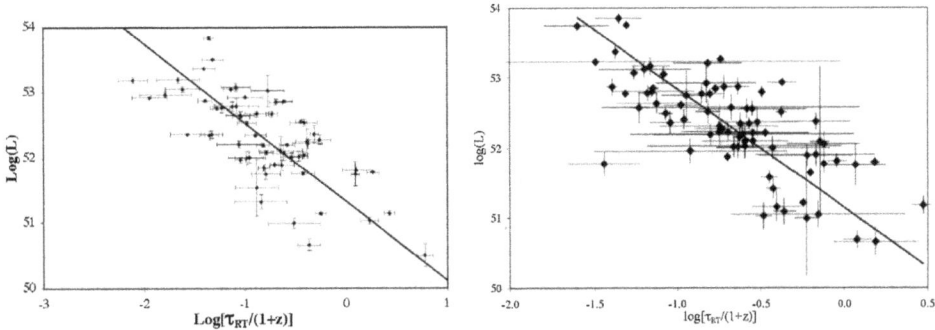

Figure 3.3. Left panel: the $\log L_{iso}$–$\log \tau_{RT}^{*}$ correlation with best-fit line and errors displayed. (Reproduced with permission from Schaefer (2007). Copyright 2007 the American Astronomical Society.) Right panel: the $\log L_{iso}$–$\log \tau_{RT}^{*}$ correlation and its best-fit line. (Reproduced with permission from Xiao and Schaefer (2009). Copyright 2009 The American Astronomical Society.)

his analysis the quickest increase in a light curve is linked to Γ. The luminosity of the burst is proportional to Γ^N with $3 < N < 5$. For this reason, the τ_{RT}–Γ and L_{iso}–Γ correlations are able to determine the L_{iso}–τ_{RT} correlation.

3.3.4 The Γ_0–E_{prompt} and Γ_0–L_{iso} correlations and their physical interpretation

Freedman and Waxman (2001) pointed out that, within the fireball scenario, if the emission reaching the observer is beamed in an angle $\simeq 1/\Gamma(t)$, the total energy E release should be interpreted as spherically symmetric, $\epsilon_e = \xi_e E/4\pi$, the energy of the fireball divided by the solid angle, where ξ_e is the electron energy fraction. Within this scenario it was found that

$$\Gamma(t) = 10.6\left(\frac{1+z}{2}\right)^{3/8}\left(\frac{E_{prompt}}{n_0}\right)^{1/8} t^{-3/8}, \qquad (3.26)$$

where E_{prompt} is in units of 10^{53} erg, n_0 is the fireball uniform ambient medium in units of cm^{-3} and t is the duration of the expansion of the fireball measured in days. As previously found by Waxman (1997), Wijers and Galama (1999) and Granot *et al* (1999) for GRB 970508, ξ_e from the afterglow observations should have a value close to equipartition, namely $\xi_e \simeq \frac{1}{3}$. A particular case is given by Wijers and Galama (1999) for GRB 971214, where $\xi_e \simeq 1$.

Later, Liang *et al* (2010) found a log Γ_0–log E_{prompt} correlation with the Lorentz factor at the beginning of the fireball emission Γ_0 (see the left panel of figure 3.4) using a dataset of 20 optical and 12 X-ray *Swift* GRBs:

$$\log\Gamma_0 = (2.26 \pm 0.03) + (0.25 \pm 0.03)\log E_{prompt}, \qquad (3.27)$$

giving $\rho = 0.89$, $P < 10^{-4}$ and $\sigma = 0.11$.

The majority of GRBs with a lower limit of Γ_0 are inside the 2σ region (see the dashed lines in the left panel of figure 3.4), while GRBs with a tentative Γ_0 computed from the afterglow or the RS peaks are usually found above the best-fit line. Smaller values of Γ_0 were in agreement with this correlation. This result occurs when Γ_0 are computed through a sample of optical afterglow light curves displaying decay since the beginning of the observation.

More recently, Ghirlanda *et al* (2011) and Lü *et al* (2012) validated the log Γ_0–log E_{prompt} correlation. They first analysed the spectra between 8 keV and 35 MeV of 13 *Fermi*/GBM SGRBs. Their findings verified the previous results.

To compute Γ_0, Lü *et al* (2012) applied three approaches to a dataset of 51 GRBs with spectroscopic redshifts. The first method requires T_{peak} of the afterglow phase as the time at which the external FS slows down (Sari and Piran 1999); the second approach assumes that the gamma-rays in the GeV band are transparent to the creation of electron–positron pairs, providing a lower constraint on Γ_0 of the emitting region (Lithwick and Sari 2001); the last method considers that an upper limit on Γ_0 can be computed from the inactivity interval between the prompt pulses (Zou and Piran 2010), in which the external shock must vanish. Employing 38 GRBs with Γ_0 obtained through the first method, they found

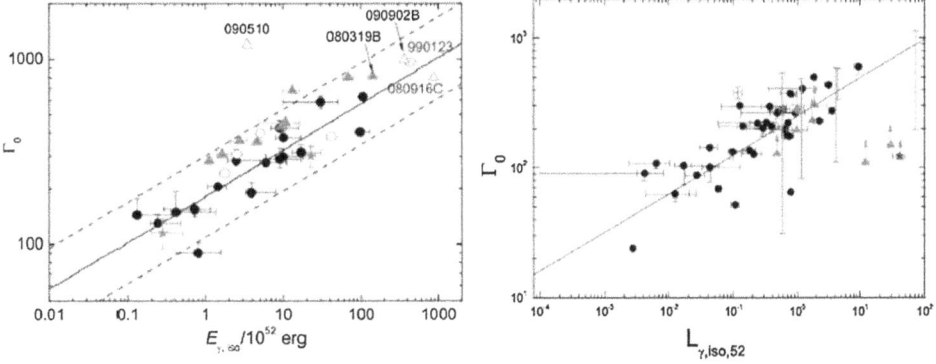

Figure 3.4. Left panel: the $\log \Gamma_0$–$\log E_{\mathrm{prompt}}$ distribution. The red stars represent GRBs 070208 and 080319C, which have a visible afterglow phase in X-rays, pink open circles indicate the GRBs with Γ_0 obtained by RS peaks or possible afterglow peaks, pink solid triangles show the GRBs with lower values of Γ_0 calculated from single power-law decay light curves and red open triangles display GRBs with strong lower values of Γ_0 for GRBs 080916C, 090902B and 090510 retrieved from the limits of opacity through detections by *Fermi*/LAT. The best-fit line is represented by the solid line. The 2σ deviation is indicated by the two dashed lines. (Reproduced with permission from Liang *et al* (2010). Copyright 2010 The American Astronomical Society.) Right panel: the $\log \Gamma_0$–$\log L_{\mathrm{iso}}$ correlation. The triangles indicate the GRBs with only lower limits and the star represents SGRB 090510. (Reproduced with permission from Lü *et al* (2012). Copyright 2012 The American Astronomical Society.)

$$\log \Gamma_0 = (1.96 \pm 0.002) + (0.29 \pm 0.002) \log E_{\mathrm{prompt}}, \qquad (3.28)$$

with $r = 0.67$ and E_{prompt} in units of 10^{52} erg. Furthermore, correcting the energy for the beaming effect, they also found a correlation between Γ_0 and L_{iso} (see the right panel of figure 3.4) given by:

$$\log \Gamma_0 = (2.40 \pm 0.002) + (0.30 \pm 0.002) \log L_{\mathrm{iso}}, \qquad (3.29)$$

with $r = 0.79$.

With regard to the physical interpretation, Liang *et al* (2010) interpreted this correlation as being due to the connection of E_{prompt} with Γ_0, stressing the validity of the fireball deceleration model. Lü *et al* (2012) claimed that a jet fuelled by neutrino-annihilation describes the correlation well. This suggests the presence of a BH spin that is not very fast and a high rate of accretion. Zhang and Pe'er (2009), Fan (2010), and Zhang and Yan (2011) proposed a model consisting of a magnetic dominated jet, because it has already been shown that magnetic fields are an important characteristic for the BH central engine models (Lei *et al* 2009).

3.3.5 Correlations between the energetics and the peak energy

The $\langle E_{\mathrm{peak}} \rangle$–$F_{peak}$ and the E_{peak}–S_{tot} correlations
Selecting 399 GRBs with $F_{\mathrm{peak}} \geqslant 1$ ph cm^{-2} s^{-1} between 50 and 300 keV, Mallozzi *et al* (1995) found a correlation between $\langle E_{\mathrm{peak}} \rangle$ and F_{peak} with $\rho = 0.90$ and $P = 0.04$. E_{peak} is given by the Comptonized photon model:

$$\frac{\mathrm{d}N}{\mathrm{d}E} = Ae^{-E(2+\beta_S)/E_{\text{peak}}}\left(\frac{E}{E_{\text{piv}}}\right)^{\beta_S}, \tag{3.30}$$

where A indicates the normalization, β_S represents the spectral index and the pivotal energy $E_{\text{piv}} = 100$ keV. Indeed, dividing the dataset into five bins of F_{peak} (see figure 3.5) it appeared that dim GRBs had a smaller $\langle E_{\text{peak}} \rangle$.

Then, adopting 1000 simulated GRBs at similar energies to those by Mallozzi *et al* (1995), Lloyd *et al* (2000a) found a significant E_{peak}–S_{tot} correlation (see the left panel of figure 3.6) given by:

$$\log E_{\text{peak}} \sim 0.29 \log S_{\text{tot}}, \tag{3.31}$$

with $\tau = 0.80$ and $P = 10^{-13}$.

Moreover, they compared the E_{peak}–S_{tot} correlation to the E_{peak}–F_{peak} correlation (see the right panel of figure 3.6). They selected a subsample composed of only the brightest GRBs, using the following selection criteria: $F_{\text{peak}} \geqslant 3$ ph cm^{-2} s^{-1}, $S_{\text{obs}} \geqslant 10^{-6}$ erg cm^{-2} and $S_{\text{tot}} \geqslant 5 \times 10^{-6}$ erg cm^{-2}. This choice guarantees that only the spectral parameters that are robust and not close to the detector threshold are taken into account in this analysis. Indeed, the sensitivity over a specific interval of energies of the instruments, in particular BATSE, and the limits on the trigger, do not allow one to avoid observational biases. However, the brightest GRBs show a weak E_{peak}–F_{peak} correlation. On the other hand, a tight E_{peak}–S_{tot} correlation was found for both the whole sample and the sample composed of the brightest GRBs,

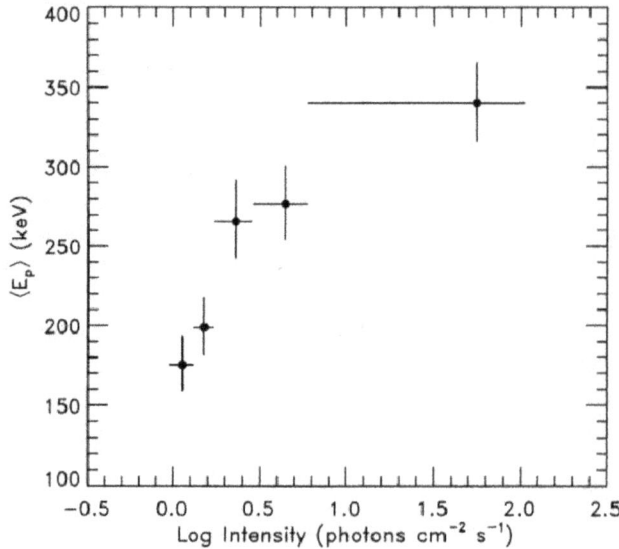

Figure 3.5. Peak energies versus intensity with displayed error bars for five GRB sets. The log intensity bin widths are indicated by the horizontal bars. (Reproduced with permission from Mallozzi *et al* (1995). Copyright 1995 The American Astronomical Society.)

for which it is simpler to treat the truncation biases and the interpretation of the cosmology is easier.

The study of the Amati (see section 3.3.5) and the Ghirlanda (see section 3.3.5) correlations is grounded on this correlation. The outcomes by Lloyd *et al* (2000a) were independent of the number of GRBs or the distribution of other parameters. However, the data results from GRBs with known redshift were in disagreement with the narrow distributions for the radiated energy or luminosity, so further investigation was needed.

Later, Goldstein *et al* (2010) found that E_{peak}/S_{tot} can be a useful estimator of the ratio between the energy at which the majority of gamma-rays are emitted and the total energy. They found out that the E_{peak}–S_{tot} correlation is important for distinguishing LGRBs from SGRBs. The most important point of this correlation is that since the energy ratio is proportional only to the square of the luminosity distance (hereafter $D_L(z, \Omega_M, \Omega_\Lambda)$), it is easy to remove the cosmological dependence from the discussed parameters. This allows the use of the energy ratio to classify GRBs.

Finally, using 51 LGRBs and 11 bright SGRBs detected by *Fermi*/GBM, Lu *et al* (2012) investigated the E_{peak}–S_{tot} correlation. They calculated the distance of the data from the best-fit line of the correlation, obtaining a value for the scatter of 0.17 ± 0.08. This outcome was already presented by Golenetskii *et al* (1983), Borgonovo and Ryde (2001), Ghirlanda *et al* (2010), Guiriec *et al* (2010) and Ghirlanda *et al* (2011).

The E_{peak}–E_{iso} correlation

Due to the limited number of GRBs observed with precise redshift, the correlation between E_{peak} and S_{tot} was discovered in the observer frame, as shown in section 3.3.5. Amati *et al* (2002) found a tighter E_{peak}–E_{iso} correlation, also known as the Amati correlation, using 12 GRBs with known redshifts detected by *BeppoSAX*. The correlation is given by:

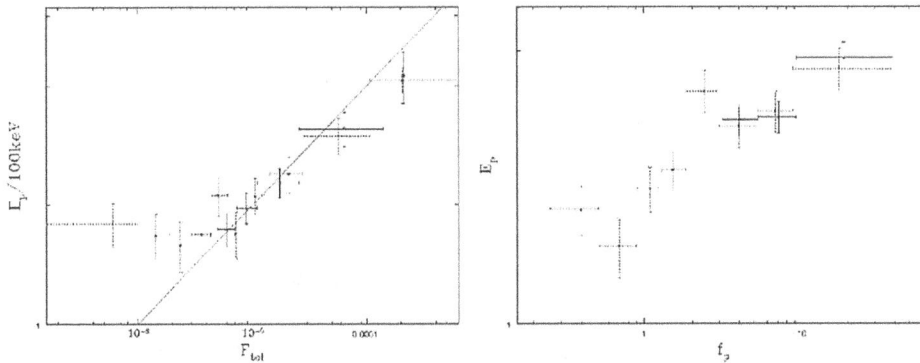

Figure 3.6. E_{peak} versus (left panel) S_{tot} and (right panel) F_{peak} distributions for the complete and subspectral samples indicated by dashed and solid points. The best-fit is represented by the solid line. (Reproduced with permission from Lloyd *et al* (2000a). Copyright 2000 The American Astronomical Society.)

$$\log E_{\text{peak}} \sim (0.52 \pm 0.06) \log E_{\text{iso}}, \tag{3.32}$$

with $r = 0.949$, $P = 0.005$ and E_{iso}:

$$E_{\text{iso}} = 4\pi D_L(z, \Omega_M, \Omega_\Lambda)^2 S_{\text{tot}}(1 + z)^{-2}. \tag{3.33}$$

Differently to Bloom *et al* (2001), Amati *et al* (2002) computed the GRB spectra in the rest-frames and obtained the radiated energy using the Band *et al* (1993) functional form between 1 and 10^4 keV. In this work a Friedman–Lemaître–Robertson–Walker cosmological model was assumed with $H_0 = 65$ kms^{-1}Mpc^{-1}, $\Omega_M = 0.3$ and $\Omega_\Lambda = 0.7$. The cosmological time dilation and spectral redshift have also been taken into consideration.

Adding 20 *BeppoSAX* GRBs with measured redshift to the sample used in Amati *et al* (2002, 2003) indicated the following correlation:

$$\log E_{\text{peak}} = (2.07 \pm 0.03) + (0.35 \pm 0.06)\log E_{\text{iso}}, \tag{3.34}$$

with $r = 0.92$, $P = 1.1 \times 10^{-8}$, E_{peak} in keV and E_{iso} in units of 10^{52} erg. The correlation is highly significant given that it has been established through a bigger sample than the one used in Amati *et al* (2002), but with a compatible correlation coefficient.

Using *HETE-2* GRBs and enlarging the sample to XRRs and XRFs, Lamb *et al* (2004) and Sakamoto *et al* (2004) obtained similar outcomes to earlier ones with a correlation spanning over three orders of magnitude in E_{peak} and five orders of magnitude in E_{iso}. These results were improved by Ghirlanda *et al* (2004b) who, with new 29 events with known redshifts, found a correlation with $r = 0.803$ and $P = 7.6 \times 10^{-7}$ (see the left panel in figure 3.8). The increase in the number of GRBs was possible due to the additional measurements of the redshift.

Ghirlanda *et al* (2005a) confirmed the E_{peak}–E_{iso} correlation obtained from 442 BATSE LGRBs for which the estimate of E_{peak} was possible and with pseudo-redshifts obtained through the L_{peak}–τ_{lag} correlation. They compared the correlation derived from this sample with that from a sample of 27 GRBs with measured spectroscopic redshifts. Due to the negligible influence of the outliers, the scatter of the data in both the correlations has compatible values, but the slope of the correlation for the 442 BATSE LGRBs is slightly smaller (0.47) than the one for the 27 GRBs with measured spectroscopic redshifts (0.56).

Amati (2006) investigated further the E_{peak}–E_{iso} correlation (see the upper left and bottom left panels of figure 3.7) by employing the following requirements: (a) 41 LGRBs/XRFs with measured z and E_{peak}, (b) 12 GRBs with uncertain z and/or E_{peak}, (c) two SGRBs with confirmed z and E_{peak}, and (d) the low-energetic GRB 980425/SN 1998bw and GRB 031203/SN 2003lw. The upper right panel of figure 3.7 presents the samples. They obtained:

$$\log E_{\text{peak}} = 1.98^{+0.05}_{-0.04} + (0.49^{+0.06}_{-0.05}) \log E_{\text{iso}}, \tag{3.35}$$

with $\rho = 0.89$, $P = 7 \times 10^{-15}$ and the same units as in equation (3.34). From this analysis it was found that this correlation can be a significant discriminant among distinct GRB types, because some of the low-energetic GRBs (980425 and possibly 031203) and SGRBs appear to not follow the $E_{\rm peak}$–$E_{\rm iso}$ correlation, while the case of GRB 060218 is consistent with the correlation. Thus, more data can shed light on this topic. The normalization of the correlation obtained by Amati (2006) is comparable to those retrieved by different space missions.

Ghirlanda *et al* (2008) found out that XRFs obey the $E_{\rm peak}$–$E_{\rm iso}$ correlation. Using 76 GRBs detected from *HETE-2*, KONUS/*Wind*, *Swift* and *Fermi*/GBM, they found the following correlation:

$$\log E_{\rm peak} \sim (0.54 \pm 0.01)\log E_{\rm iso}. \qquad (3.36)$$

This result suggested a strong correlation without additional outliers (except for the well known outliers GRB 980425 and GRB 031203).

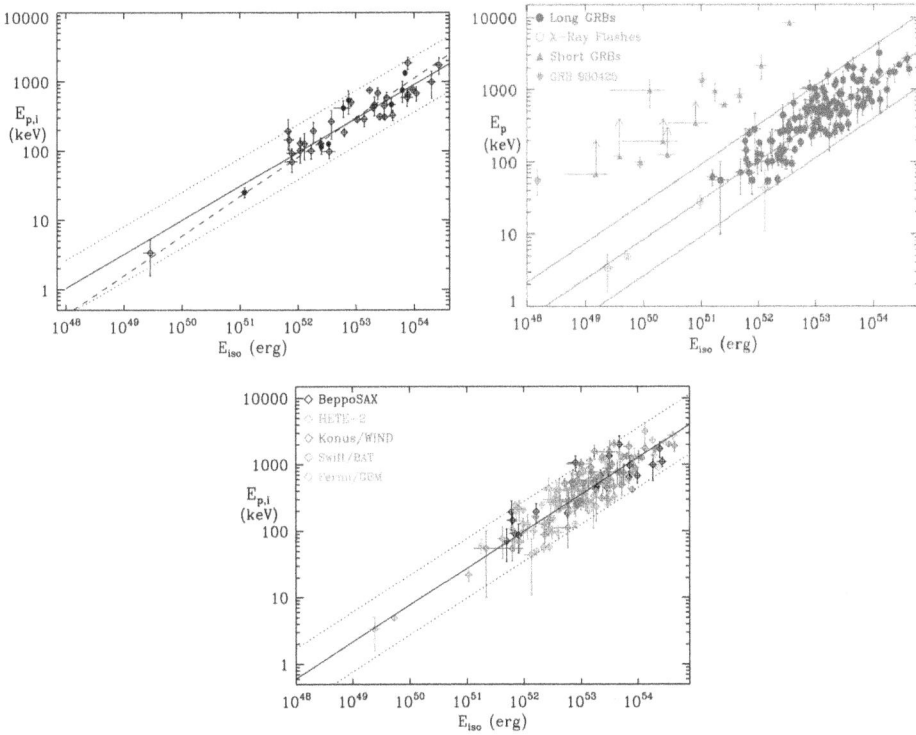

Figure 3.7. Upper left panel: the $E_{\rm peak}$–$E_{\rm iso}$ distribution for 41 GRBs/XRFs. *Swift* GRBs are represented by the filled circles. The best-fit line is indicated by the solid line, while the best-fit line without the sample variance is displayed by the dashed line. (Reproduced from Amati (2006).) Upper right panel: the $E_{\rm peak}$–$E_{\rm iso}$ correlation with the best-fit line and the $\pm 2\,\sigma$ confidence area for LGRBs and XRFs displayed. (Reproduced with permission from Amati (2012). Copyright 2012 World Scientific Publishing.) Bottom left panel: the $E_{\rm peak}$–$E_{\rm iso}$ correlation for 12 GRBs, the sub-energetic event GRB 980425, and for the SGRBs 050709 and 051221. *Swift* GRBs are indicated by filled circles. The best-fit line is represented by the solid line, while the best-fit line calculated without the sample variance is displayed from the dashed line. (Reproduced from Amati (2006).)

Later, using 95 *Fermi* GRBs with known z, Amati *et al* (2009) found the following E_{peak}–E_{iso} correlation:

$$\log E_{peak} \sim 0.57 \log E_{iso}, \qquad (3.37)$$

with $\rho = 0.88$ and $P < 10^{-3}$. Analysing the highly energetic LGRBs 080916C and 090323 and the SGRB 090510, Amati *et al* (2009) pointed out that LGRBs, whether they are very energetic or less luminous, still follow the E_{peak}–E_{iso} correlation, while SGRBs do not follow the correlation, as in the case of GRB 090510, again independently of their luminosities. They concluded that in the brightest and most energetic LGRBs the emission mechanism should be the same as for other long events with an average luminosity and XRFs, because they all satisfy the Amati correlation.

Using a broadened sample of 120 GRBs, Amati (2012) updated the work by Amati *et al* (2008) (see the upper right panel of figure 3.7) pointing out the correlation

$$\log E_{peak} = 2 + 0.5 \log E_{iso}, \qquad (3.38)$$

with the same units as in equations (3.34) and (3.35). Employing a set of 153 GRBs with known z, E_{peak}, E_{iso} and T_{90} detected by several instruments, Qin and Chen (2013) studied the discrepancy between measured E_{peak} and those derived by the best-fit line of the Amati correlation. They discovered an explicit bimodality: the Amati GRBs, obeying the Amati correlation, and the non-Amati GRBs, not obeying it. In the case of Amati type GRBs it was obtained that

$$\log E_{peak} = (2.06 \pm 0.16) + (0.51 \pm 0.12)\log E_{iso}, \qquad (3.39)$$

with $r = 0.83$ and $P < 10^{-36}$, instead adopting the set of non-Amati GRBs:

$$\log E_{peak} = (3.16 \pm 0.65) + (0.39 \pm 0.33)\log E_{iso}, \qquad (3.40)$$

with $r = 0.91$, $P < 10^{-7}$, E_{peak} in keV and E_{iso} in units of 10^{52} erg. Furthermore, they claimed that Amati GRBs are connected to energetic LGRBs, while non-Amati GRBs are in the majority of cases associated with SGRBs. The two groups of GRBs are noticeably discriminated, thus these two types of GRBs can be catalogued without difficulty.

Selecting 43 *Fermi* GRBs (with measured redshifts) based on specific criteria regarding the duration and spectral indices, Heussaff *et al* (2013) found the correlation:

$$\log E_{peak} = 2.07 + 0.49 \log E_{iso}, \qquad (3.41)$$

with $\rho = 0.70$, $P = 1.7 \times 10^{-7}$. The correlation has been computed with units analogous to the earlier Amati correlations.

Basak and Rao (2012) claimed that a time-resolved Amati correlation is also verified in each GRB with normalization and slope compatible with those obtained with time-averaged spectra. Actually, this correlation appears even stronger than the time-integrated correlation (Basak and Rao 2013). Time-resolved E_{peak} and E_{iso}

were computed at distinct moments of the prompt phase (see also Ghirlanda *et al* 2010, Lu *et al* 2012, Frontera *et al* 2012 and section 3.3.6).

The E_{peak}–E_γ correlation
Ghirlanda *et al* (2004b) determined for the first time the E_{peak}–E_γ correlation where E_γ is the energy corrected for the beaming factor:

$$E_\gamma = (1 - \cos\theta_{jet}) \times 4\pi \times D_L^2(z, \Omega_M, h) \times S_{\gamma,prompt}/(1 + z)^2. \qquad (3.42)$$

This correlation is also called the Ghirlanda correlation employing a set of 40 GRBs with measured z and E_{peak}. To compute E_γ using E_{iso}, θ_{jet}, which, however, is observed only for less than half of the bursts, is needed. For this parameter, theoretical assumptions need to be made regarding the density medium, the radiative efficiency and the energy, see equation (2.6) in section 2.3. They found a correlation with slope 0.7, $\rho = 0.88$ and $P = 2.7 \times 10^{-8}$.

In contrast to Ghirlanda *et al* (2004b), Liang and Zhang (2005) took into account a sample composed of 15 GRBs with known z, E_{peak} and a purely phenomenological time T_{break} of the afterglow phase in the optical band. Thus, circumventing any hypotheses on the theoretical model regarding T_{break}, they found the following correlation:

$$\log E_\gamma = (0.85 \pm 0.21) + (1.94 \pm 0.17)\log E_{peak}^* - (1.24 \pm 0.23)\log T_{break}^*, \qquad (3.43)$$

where E_γ is in units of 10^{52} erg, E_{peak}^* in units of 100 keV, T_{break}^* is measured in days, $\rho = 0.96$ and $P < 10^{-4}$.

From their analysis, Nava *et al* (2006) claimed that the Ghirlanda correlation for a wind-like circumburst medium is as tight as the one for a homogeneous medium. Drawing a comparison between the observed and the comoving frame correlations, the wind-like Ghirlanda correlation also remains linear in the comoving frame, regardless of the Γ values. While, for the homogeneous density medium case, a significant correlation between Γ and the total energy should exist, thus further constraining the emission models of prompt radiation. A dataset of 18 GRBs with firm z, E_{peak} and T_{break} was used by Nava *et al* (2006) for the homogeneous density case, obtaining:

$$\log \frac{E_{peak}^*}{100\text{ keV}} = 0.45^{+0.02}_{-0.03} + (0.69 \pm 0.04) \log \frac{E_\gamma}{2.72 \times 10^{52}\text{ erg}}, \qquad (3.44)$$

with $\rho = 0.93$ and $P = 2.3 \times 10^{-8}$. For the wind case, they found:

$$\log \frac{E_{peak}^*}{100\text{ keV}} = 0.48^{+0.02}_{-0.03} + (1.03 \pm 0.06) \log \frac{E_\gamma}{2.2 \times 10^{50}\text{ erg}}, \qquad (3.45)$$

with $\rho = 0.92$ and $P = 6.9 \times 10^{-8}$.

Ghirlanda *et al* (2007), employing a sample of 33 GRBs (16 *Swift* GRBs with measured z and E_{peak}, and 17 GRBs detected by pre-*Swift* missions), investigated the E_{peak}–E_γ correlation. To calculate the T_{break} they suggested that:

1. The jet break has to be observed in the optical wavelengths.
2. The optical light curve should last long after T_{break}.
3. The flux from a plausible SN and the host galaxy emission should not be taken into account.
4. The optical break should not be influenced by the frequency, and an X-ray break at the same time is not needed.
5. T_{break} should not coincide with $T_{X,a}$ from W07, otherwise the characteristic influencing the X-ray flux is also affecting the optical flux.

From all these assumptions, the final set dropped down to 16 GRBs which obey the following correlation:

$$\log \frac{E_{peak}}{100 \text{ keV}} = (0.48 \pm 0.02) + (0.70 \pm 0.04) \log \frac{E_\gamma}{4.4 \times 10^{50} \text{ erg}}. \quad (3.46)$$

In this dataset there were no outliers, thus the decreased value of the scatter in this correlation strengthened the possibility to conceive GRBs as standardizable candles.

Physical interpretation of the energetics versus peak energy correlations
Under the assumption of a synchrotron radiation process from internal and external shocks, Lloyd *et al* (2000a) enquired into the physical explanation of the E_{peak}–S_{tot} correlation. The synchrotron emission from electrons with a power-law distribution and Γ larger than some minimum threshold value, Γ_m, is able to explain the correlation. Moreover, the internal shock model explains the E_{peak}–S_{tot} correlation and the radiated energy better than the external shock model.

Lloyd-Ronning and Petrosian (2002) stated that the GRB particle acceleration problem needs further investigation. The common hypothesis is that the repeated scatterings through the (internal) shocks accelerate the emitted particles. The recurrent shocks are generated by particles with a power-law distribution and a large photon index. The link between E_{peak} and the photon flux can be clarified by the fluctuation in the magnetic field or electron energy. In conclusion, it was found that in almost all GRBs the particle acceleration is not an isotropic process, but happens along the magnetic field lines.

In their analysis Amati *et al* (2002) validated the result that the $\log E_{peak}$ ~ 0.5 $\log E_{iso}$ correlation is obtained considering an optically thin synchrotron shock model, as already shown by Lloyd *et al* (2000b). In the framework of this model electrons follow the $N(\Gamma) = N_0 \Gamma^{-p}$ distribution for $\Gamma > \Gamma_m$, with Γ_m, GRB duration and N_0, the normalization, which are constant factors for each GRBs. A caveat must be posed regarding these assumptions, because each GRB has a different duration and E_{iso} might be smaller when the emission is beamed.

The effect of this correlation on the theoretical interpretation of the prompt emission and on the probable union of the two classes of GRBs and XRFs was presented by Amati (2006). As shown by Zhang and Mészáros (2002) and Ghirlanda *et al* (2013), this correlation can also be employed for verifying the GRB synthesis models.

E_{peak} and E_{iso} are dependent on Γ and the E_{peak}–E_{iso} correlation is a useful tool to link the parameters of the synchrotron shock model. Indeed, Zhang and Mészáros (2002) and Rees and Mészáros (2005) concluded that, for a power-law electron distribution generated from an internal shock in a fireball with velocity Γ, the peak energy E_{peak}^{*} is

$$\log E_{peak}^{*} \sim -2 \log \Gamma + 0.5 \log L - \log t_{\nu}, \tag{3.47}$$

where L is the total fireball luminosity and t_{ν} the variability timescale. To retrieve the E_{peak}–E_{iso} correlation a complex assumption has to be made: Γ and t_{ν} should be similar for each GRB. This assumption causes problems when $L \propto \Gamma^{N}$, with N between 2 and 3 (Zhang and Mészáros 2002, Schaefer 2003b, Ramirez-Ruiz 2005). A possible interpretation could be given by the fact that the GRB prompt emission is affected by direct or Comptonized thermal radiation produced by the photospheric region of the fireball (Zhang and Mészáros 2002, Ramirez-Ruiz 2005, Ryde 2005, Rees and Mészáros 2005, Beloborodov 2010, Guiriec *et al* 2011, Hascoët *et al* 2013, Guiriec *et al* 2013, Vurm and Beloborodov 2016, Guiriec *et al* 2015a, 2015b). This explanation is particularly suitable for very energetic (Frontera *et al* 2000, Preece *et al* 2000, Ghirlanda *et al* 2003) and flat average spectra. In these circumstances, E_{peak} is a function of the peak temperature, $T_{bb,peak}$, of the photons distributed as in a blackbody, and for this reason it is related to the luminosity or emitted energy. In the case of Comptonized radiation from the photosphere $\log E_{peak}$ can be written as:

$$\log E_{peak} \sim \log \Gamma + \log T_{bb,peak} \sim 2 \log \Gamma - 0.25 \log L \tag{3.48}$$

or

$$\log E_{peak} \sim \log \Gamma + \log T_{bb,peak} \sim -0.5 \log r_0 + 0.25 \log L, \tag{3.49}$$

where r_0 is the distance between the region of energy emission and the central engine, such that $\Gamma \simeq r/r_0$ increases up to some saturation radius r_s (Rees and Mészáros 2005). Rees and Mészáros (2005) claimed that in this case the E_{peak}–E_{iso} correlation can be found just below the photosphere, although it would be a function of an uncertain amount of unknown parameters.

The non-thermal synchrotron emission model can have an important role in explaining the E_{peak}–E_{iso} correlation for highly energetic GRBs (i.e. $E_{iso} \approx 10^{55}$ erg). As shown by Lloyd *et al* (2000b) and Zhang and Mészáros (2002) this can be achieved in two ways: (a) supposing the minimum Γ and the normalization of the electron distribution approximated as a power-law, constant in each GRB and (b) by bounding the slope of the correlation between Γ and the luminosity.

Panaitescu (2009), from the study of 76 GRBs with measured redshifts, concluded that the $\log E_{peak} \sim 0.5 \log E_{iso}$ correlation for LGRBs is generated from the external shock interacting with an external medium which is radially stratified, for example with a particle density distribution which does not follow the usual R^{-2} trend, with R the distance from the produced GRB radiation.

To investigate the connection between the E_{peak}–E_{iso} correlation and the internal shock scenario, Mochkovitch and Nava (2015) simulated GRB sets with different model parameter distributions (e.g. the radiated power in the relativistic emission and Γ). Then, they compared the E_{peak}–E_{iso} correlations obtained through the simulated dataset with the observed correlation recovered using 58 *Swift* GRBs with $F_{peak} > 2.6$ ph cm^{-2} s^{-1} between 15 and 150 keV. An agreement between observations and simulations was obtained only with the following assumptions:

1. The diffused energy should be emitted in a small amount of electrons.
2. The difference between the highest and the lowest Γ should be irrelevant.
3. If $\bar{\Gamma} \propto \dot{E}^{1/2}$ (with $\dot{E}$ the rate of injected energy in the relativistic jet), the E_{peak}–E_{iso} correlation is not recovered and E_{peak} decreases with greater values of E_{iso}. However, the E_{peak}–E_{iso} correlation can be found if $\bar{\Gamma} \propto \dot{E}^{1/2}$ is a lower limit for a specific $\dot{E}$.
4. The E_{peak}–E_{iso} correlation is retrieved when the amplitude of the Γ variability is related to $\bar{\Gamma}$. $\bar{\Gamma}$ is the average Lorentz factor (for equations see equation 1 of Mochkovitch and Nava 2015).

The Ghirlanda correlation (Ghirlanda *et al* 2004b) has been considered invariant under the transition from the rest-frame to the comoving frame, assuming that θ_{jet} is along the line of sight. The number of emitted photons for each GRB is similar and around 10^{57}. This aspect could be significant to understand the GRB dynamics and the radiation processes (see also the right panel of figure 3.8).

Collazzi *et al* (2011) investigated the mean E_{peak}^*, finding that it is close to the electron rest-mass energy value, 511 keV. This result suggests that the shape of the E_{peak} distribution is not caused by selection biases only. However, this effect is not explained by any known mechanism, but the almost steady temperature indicates the necessity of some mechanism holding the temperature at a constant value, such as electron–positron annihilation. From their analysis of a simulated dataset, Ghirlanda *et al* (2013) found that the correlation of Γ and θ_{jet} to have the highest agreement between simulations and observations is given by $\theta_{jet,max}^{2.5} \Gamma_{max} = $ const. Using a set of $\approx$30 GRBs with computed θ_{jet} or Γ a correlation between E_γ and E_{peak} was found:

$$\log E_{peak} \sim \log \frac{E_\gamma}{5 - 2\beta_0}. \tag{3.50}$$

Even if the Γ and θ_{jet} estimations are based on incomplete sets and can be affected by selection effects, Ghirlanda *et al* (2013) concluded that the higher the value of Γ, the smaller its θ_{jet}, namely the faster a GRB, the narrower its jet. Ghirlanda *et al* (2013) claimed that just $\approx 6\%$ of the on-axis GRBs should have $\sin \theta_{jet} < 1/\Gamma$, not displaying any jet break feature in the afterglow light curve. Most importantly

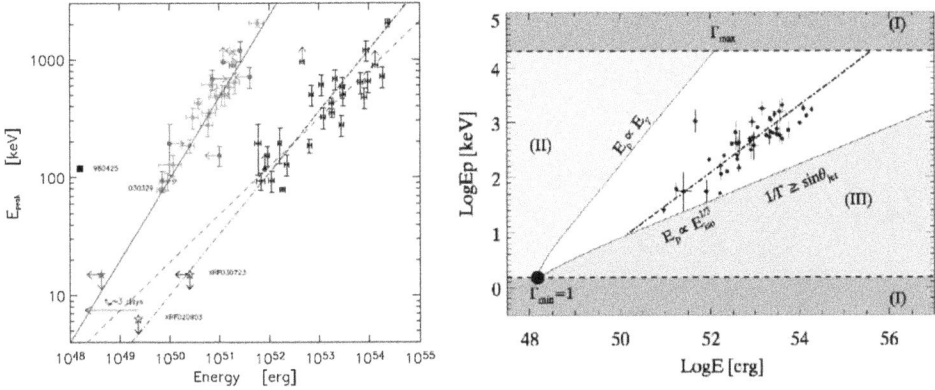

Figure 3.8. Left panel: the E^*_{peak}–E_γ correlation for GRBs with measured redshift. E_γ for GRBs where a jet break was detected is indicated by filled circles. Lower/upper limits are represented by grey symbols. The best-fit is displayed by the solid line, while open circles indicate E_{iso} for the GRBs. The best-fit to these points is given by the dashed line and the correlation shown by Amati *et al* (2002) is represented by the dash-dotted line. (Reproduced with permission from Ghirlanda *et al* (2004b). Copyright 2004 The American Astronomical Society.) Right panel: rest-frame distribution of the Ghirlanda correlation. The large black dot represents the simulated GRBs with $E^*_{\text{peak}} = 1.5$ keV and $E^*_\gamma = 1.5 \times 10^{48}$ erg. Region I, II, and III are not allowed. Region I is forbidden because $\Gamma > 1$ but less than 8000, region II because $\theta_{\text{jet}} \leqslant 90°$, and region III because in the case Γ is tiny, the area of the beaming cone appears wider than the one individuated by the jet. The black dots represent the *Swift* GRBs of the sample. The dot-dashed line displays the fit to the *Swift* sample. (Reproduced from Ghirlanda *et al* (2013). By permission of Oxford University Press on behalf of the Royal Astronomical Society.)

they concluded that the local rate of GRBs is $\approx 0.3\%$ of the local SNe Ib/c, and $\approx 4.3\%$ of the local hypernovae. For those rates see also section 6.1.1.

3.3.6 Correlations between the luminosity and the peak energy

The L_{iso}–E_{peak} correlation
Schaefer (2003a) was the first to discover the L_{iso}–E_{peak} correlation. For 20 GRBs with luminosities based on optically known redshift (Amati *et al* 2002, Schaefer 2003a) it was obtained that (see figure 3.9)

$$\log E_{\text{peak}} \sim (0.38 \pm 0.11)\log L_{\text{iso}}, \tag{3.51}$$

with $r = 0.90$ and $P = 3 \times 10^{-8}$, while for 84 GRBs with computed E_{peak} from the BATSE catalogue (Schaefer *et al* 2001) the correlation was given by

$$\log E_{\text{peak}} \sim (0.36 \pm 0.03)\log L_{\text{iso}}. \tag{3.52}$$

Given that both E_{peak} and L_{iso} are functions of Γ, their common dependence on Γ simply relates these two physical quantities. Assuming that the luminosity is proportional to Γ^N and E_{peak} is proportional to Γ^M, $\log E_{\text{peak}}$ will be a function of $(M + 1)/N \times \log L_{\text{iso}}$.

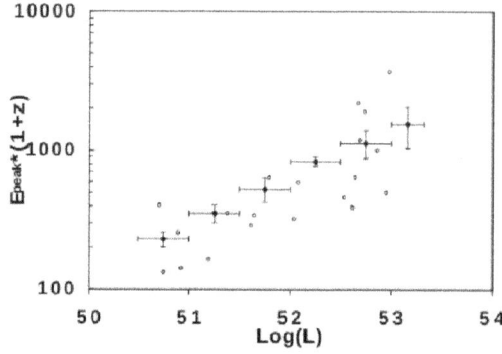

Figure 3.9. Fit of the log L_{iso}–log E_{peak} correlation. The open circles represent 20 GRBs of the sample with known redshift and the filled diamonds indicate 84 GRBs with luminosity (and then redshift) computed using spectral lag and variability. (Reproduced with permission from Schaefer (2003b). Copyright 2003 The American Astronomical Society.)

Investigating the time-resolved prompt emission spectra between 2 and 2000 keV of nine GRBs observed simultaneously with the Wide Field Camera and the BATSE instrument, Frontera *et al* (2012) obtained the correlation (see figure 3.10)

$$\log E^*_{\mathrm{peak}} \sim (0.66 \pm 0.03)\log L_{\mathrm{iso}}, \qquad (3.53)$$

with $\rho = 0.94$ and $P = 1.57 \times 10^{-13}$.

Additionally, employing 46 *Swift* GRBs with known z and E_{peak}, Nava *et al* (2012) found a significant L_{iso}–E_{peak} correlation represented by

$$\log E^*_{\mathrm{peak}} = -(25.33 \pm 3.26) + (0.53 \pm 0.06)\log L_{\mathrm{iso}}, \qquad (3.54)$$

with $\rho = 0.65$, $P = 10^{-6}$, E_{peak} in keV and L_{iso} is in units of 10^{51} erg s^{-1}. Nevertheless, they showed that, even using GRBs with no estimation or just an upper or lower limit on z and E_{peak}, this L_{iso}–E_{peak} is still valid.

The L$_{peak}$–E$_{peak}$ *correlation*
Even replacing E_{iso} with L_{iso} or L_{peak}, a correlation is still present. The Yonetoku correlation (Yonetoku *et al* (2004), see the left panel of figure 3.11), between E_{peak} and L_{peak}, was recovered with a sample of 11 BATSE GRBs with measured redshifts and *BeppoSAX* GRBs from (Amati *et al* 2002). Considering L_{peak} of the burst instead of L_{iso}, this correlation appears more significant than other correlations between prompt phase parameters:

$$\log L_{\mathrm{peak}} \sim (2.0 \pm 0.2)\log E^*_{\mathrm{peak}}, \qquad (3.55)$$

with $r = 0.958$, $P = 5.31 \times 10^{-9}$, and the uncertainties are 1σ errors. Indeed, this correlation, compatible with the standard synchrotron model (Zhang and Mészáros 2002, Lloyd *et al* 2000b), has been employed to compute the pseudo-redshifts of 689

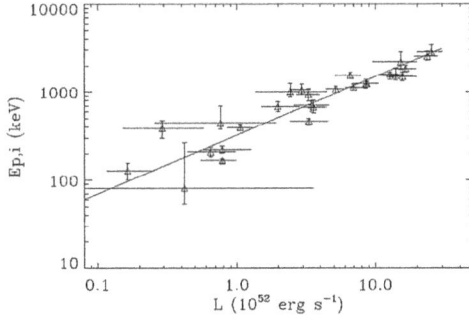

Figure 3.10. The E_{peak}^*–L_{iso} correlation obtained from data for GRBs 990123 and 990510 computed at different time intervals. The solid line displays the best-fit power-law. (Reproduced with permission from Frontera *et al* (2012). Copyright 2012 The American Astronomical Society.)

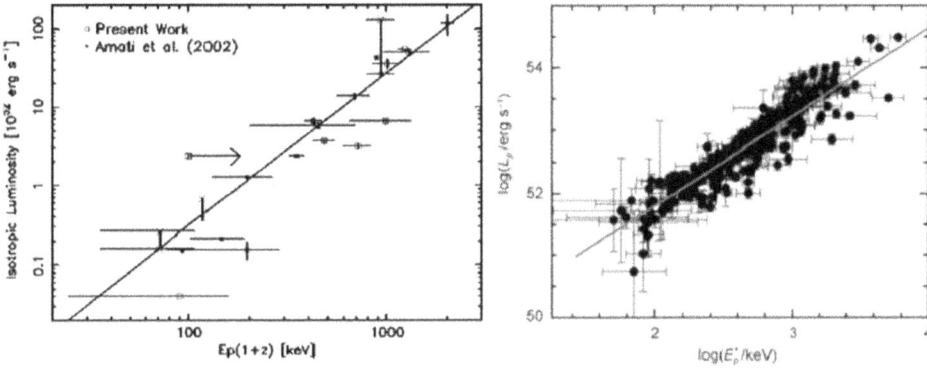

Figure 3.11. Left panel: the $\log L_{\text{peak}}$–$\log E_{\text{peak}}$ correlation with BATSE data represented by the open squares and the *BeppoSAX* GRBs indicated by the filled squares and the cross points. The best-fit line is displayed by the solid line. (Reproduced with permission from Yonetoku *et al* (2004). Copyright 2004 The American Astronomical Society.) Right panel: the $\log L_{\text{peak}}$–$\log E_{\text{peak}}$ correlation using a sample of 276 GRB time-resolved spectra within the decay pulses. The best-fit to the set is indicated by the solid line. (Reproduced with permission from Lu and Liang (2010). Copyright 2010 Springer.)

BATSE LGRBs without measured distances and to calculate the formation rate versus z.

From a sample of 36 bright BATSE SGRBs, Ghirlanda *et al* (2004a) chose the GRBs with F_{peak}, measured on the 64 ms timescale and in the energy band between 50 and 300 keV, larger than 10 ph cm^{-2} s^{-1} and with a significant signal to noise ratio (S/N), obtaining in the end a dataset of 28 SGRBs. From the analysis of this sample, it was found that SGRBs follow the L_{peak}–E_{peak}^* correlation, but not the Amati correlation. So, the pseudo-redshifts for SGRBs were calculated using the Yonetoku correlation and, as a result, their distribution resembles the one for LGRBs, but centred at a smaller redshift. Later, Yonetoku *et al* (2010) with a sample of 101 GRBs with known redshifts, reported F_{peak}, and observed by KONUS, *Swift*, HXD-WAM and *RHESSI* changed the parameter values of the correlation as follows:

$$\log L_{\text{peak}} = (52.43 \pm 0.037) + (1.60 \pm 0.082)\log E^*_{\text{peak}}, \tag{3.56}$$

with $r = 0.889$, $P = 2.18 \times 10^{-35}$, L_{peak} expressed in erg s^{-1} and E^*_{peak} in units of 355 keV. In conclusion, it was found that this correlation is intrinsic to GRBs, but the selection effects due to the detector threshold severely influence it.

With time-resolved spectral data for 30 pulses in 27 bright BATSE GRBs, Lu and Liang (2010) studied the decaying parts of the pulses (see the right panel of figure 3.11) obtaining:

$$\log L_{\text{peak}} \sim (1.42 \pm 0.03)\log E^*_{\text{peak}}, \tag{3.57}$$

with $r = 0.91$ and $P < 10^{-4}$. The statistical or observational effects are not alone sufficient to justify the wide scatter of the power-law index. Thus, this may be an intrinsic characteristic, suggesting that there is no common L_{peak}–E_{peak} correlation expected for all GRB pulses. Considering the *Fermi* observations deviating from the Band function (Abdo *et al* 2009a, Guiriec *et al* 2010, Ackermann *et al* 2010, 2011, 2013; see also Lin *et al* 2016b), Guiriec *et al* (2013, Guiriec *et al* 2015a, 2015b, Guiriec *et al* 2016) suggested that the GRB spectra should be described not only by the Band function, but also by blackbody (thermal) and power-law (non-thermal) components. Synchrotron radiation from the jet particles interpreted the non-thermal component well, while the radiation generated by the jet photosphere described the thermal component. Most likely, the inverse Compton process produced the power-law component, but the results suggest a general L_{peak}–E^*_{peak} correlation due to the non-thermal components.

In a sample of 13 SGRB candidates, Tsutsui *et al* (2013) chose eight well-defined SGRBs (defined by a scatter larger than $3\sigma_{\text{int}}$ around the Amati correlation for LGRBs) in order to study the L_{peak}–E_{peak} correlation. It was found that

$$\log L_{\text{peak}} = (52.29 \pm 0.066) + (1.59 \pm 0.11)\log E^*_{\text{peak}}, \tag{3.58}$$

with $r = 0.98$, $P = 1.5 \times 10^{-5}$, E^*_{peak} in units of 774.5 keV from the time-integrated spectrum and L_{peak} in erg s^{-1} integrated for 64 ms. Employing this correlation on 71 luminous BATSE SGRBs allowed the computation of their pseudo-redshifts, which were spread in the range $z \in [0.097, 2.258]$, with a mean value $\langle z \rangle = 1.05$, smaller than $\langle z \rangle = 2.2$ for LGRBs. In conclusion, Yonetoku *et al* (2014), considering a sample of 72 SGRBs with well known spectral characteristics detected by BATSE, used the L_{peak}–E_{peak} correlation for SGRBs obtained by Tsutsui *et al* (2013) to calculate their pseudo-redshifts and luminosities. The measured redshift distribution for $z \leqslant 1$ was comparable to the one for 22 *Swift* SGRBs, confirming the accuracy of the redshift estimation through the E^*_{peak}–L_{peak} correlation.

Physical interpretation of the luminosity versus peak energy correlations
Schaefer *et al* (2001) and Schaefer (2003b) showed that E_{peak} and L_{iso} are linked due to their common proportionality to the factor Γ. As claimed by Lloyd-Ronning and Ramirez-Ruiz (2002), the L_{iso}–E_{peak} correlation could help to explain the structure

of the ultra-relativistic jet, the shock acceleration and the production of the magnetic field. However, the small number of SGRBs present in the datasets allows the correlation to be used for LGRBs only.

Even if Schaefer *et al* (2001) and Schaefer (2003b) concluded that E_{peak} is the same for all GRBs at redshifts $z \geqslant 5$, the use of GRBs as standard candles is still an open issue. The most important result would be to find common features in some GRB quantities, regardless of the complexity and differences in each GRB light curve, in order to use GRBs as cosmological tools (Wang *et al* 2015).

Introducing the quantity $\omega = (L_{iso}/10^{52} \text{ erg s}^{-1})^{0.5}/(E_{peak}/200 \text{ keV})$, Liang *et al* (2004) constrained ω to the values $\simeq 0.1-1$ and investigated the consequences of the $E_{peak}-L_{iso}$ correlation within the fireball model. Other parameters were discussed, such as the combined internal shock parameter, ζ_i, for both the internal and external shock scenarios. With the hypothesis of uncorrelated parameters, they concluded that the production of prompt gamma-rays within internal shocks dominated by kinetic energy fulfils the requirements of the standard internal shock model. The same is valid for the gamma-rays generated from external shocks subjected to magnetic dissipation, namely both models give a valid explanation of the $L_{iso} \propto E_{peak}^2$ correlation and ω.

Another interpretation of this correlation was given by Mendoza *et al* (2009). They considered a source ejecting matter in a particular x direction with a speed $v(t)$ and a rate of mass ejection $\dot{m}(t)$, both dependent on time t as computed from the source of the outflow. Then, they investigated the uniform release of mass and, assuming simple periodic oscillations of the particle velocity (an ordinary hypothesis in the internal shock model scenario), they computed the luminosity.

In the photospheric context, Ito *et al* (2013) proposed that if a velocity shear with a significant change in Γ is observed at the edge of the spine and the sheath region, the high-energy component of the GRB photon spectrum is described by a Fermi-like acceleration mechanism. This acceleration phenomenon also explains the power-law slope above the thermal-like peak bump visible in a few GRBs (090510, 090902B, 090926A). Additionally, they claimed that time-integrated spectra can recover the low-energy part of the GRB spectrum in the case the time evolution of the jet is taken into account. For the Yonetoku correlation, Yonetoku *et al* (2004) claimed that the luminosity evolution of GRBs may be linked to the evolution of GRB progenitor mass or to the jet evolution. To investigate the evolution of the jet opening angle, two cases were taken into account: the maximum jet opening angle drops or the total jet energy rises. However, in the first scenario, the GFR is underestimated due to the fact that the probability to detect GRBs at high redshift would be reduced. In this case the evolution of the ratio of the GFR to the SFR would be faster. Instead in the second scenario, the evaluation of the GFR appears reliable.

Finally, to explain the $L_{iso}-E_{peak}$ correlation, Frontera *et al* (2016) started from the model developed by Titarchuk *et al* (2012). This model is characterized by an expanding plasma shell originating from a star explosion and soft photons. Through this model, Frontera *et al* (2016) concluded that, if $\tau_{\gamma\gamma} \gg 1$, see section 2.1, the $\log L_{\text{iso}}$–$\log E_{\text{peak}}$ correlation has a slope of 0.5, thus implying the physical ground for the Amati correlation (see section 3.3.5).

3.3.7 Comparisons between E_{peak}–E_{iso} and E_{peak}–L_{peak} correlation

Here, the E_{peak}–E_{iso} correlation and the E_{peak}–L_{peak} correlation are examined. With this aim, Ghirlanda et al (2005b) determined the E_{peak}–L_{peak} correlation, using 22 GRBs with measured z, with a slope of 0.51. This is comparable to that found by Yonetoku et al (2004) using 12 GRBs, but it presents a larger scatter than the one found by Yonetoku et al (2004).

Employing 33 low-redshift GRBs with $z \leqslant 1.6$, Tsutsui et al (2009a) studied these two correlations, finding in both of them a significant correlation coefficient together with a relevant scatter. Additionally, a partial linear correlation degree, which represents the measure of the correlation between two quantities, was $\rho_{L_{\text{peak}}, E_{\text{iso}}, E_{\text{peak}}} = 0.38$. To attenuate the scatter of the Yonetoku correlation, a parameter $T_L = E_{\text{iso}}/L_{\text{peak}}$ was inserted by Tsutsui et al (2009a) as a third parameter, giving a new correlation:

$$\log L_{\text{peak}} = (-3.87 \pm 0.19) + (1.82 \pm 0.08)\log E_{\text{peak}} - (0.34 \pm 0.09)\log T_L, \quad (3.59)$$

with $r = 0.94$, $P = 10^{-10}$, L_{peak} in units of 10^{52} erg s^{-1}, E_{peak} in keV and T_L in seconds. Through this correlation the systematic errors were diminished by 40%, and the plane displayed by this correlation might be a 'fundamental plane' for GRBs observed during their prompt emission phase.

Using the database developed by Yonetoku et al (2010) (composed of 109 GRBs with known redshifts, E_{peak}, L_{peak} and E_{iso}), Tsutsui et al (2010) analysed the correlations between E_{peak}, L_{peak} and E_{iso}. In this sample GRBs were separated into two groups: the gold and the bronze datasets. The first is composed of GRBs with E_{peak} given by the Band function with four free parameters. The latter is composed of GRBs with such a poor energy spectra that E_{peak} must be given either by the Band function with only three free parameters (instead of four, for example with one fixed spectral index) or by the three free parameters of the cut-off power-law (CPL) model. E_{peak} for the GRBs in the bronze dataset had a larger value than that estimated for the GRBs in the gold dataset, suggesting that the quality of the dataset affected the scatter of the correlations between E_{peak}, L_{peak} and E_{iso}.

The presence of GRB 060218 in the LGRB sample induces a difference between the E_{peak}–L_{peak} correlation found by Ghirlanda et al (2010) and the one analysed by Yonetoku et al (2010). For Ghirlanda et al (2010) GRB 060218 was an ordinary LGRB, while for Yonetoku et al (2010) this GRB was removed from the sample, since it was an outlier situated at more than 8σ from the L_{peak}–E_{peak} correlation. This GRB removal gave as a result a much steeper best-fit line.

Employing 13 highly energetic *Fermi* GRBs up to July 2009, and with measured redshift, Ghirlanda et al (2010) found a significant correlation:

$$\log E_{\text{peak}}^* \sim 0.4 \log L_{\text{iso}}, \tag{3.60}$$

with a scatter of $\sigma = 0.26$. Likewise, for E_{peak}^* and E_{iso}:

$$\log E_{\text{peak}}^* \sim 0.5 \log E_{\text{iso}}. \tag{3.61}$$

The time-integrated spectra of eight *Fermi* GRBs with known redshift were compatible with both the $E_{\text{peak}}-E_{\text{iso}}$ and the $E_{\text{peak}}-L_{\text{iso}}$ correlations obtained from a sample of 100 GRBs observed by missions previous to *Fermi*.

In the framework of these two correlations, eight SGRBs were employed by Tsutsui *et al* (2013) to verify whether the $E_{\text{peak}}-E_{\text{iso}}$ and $E_{\text{peak}}-L_{\text{peak}}$ correlations are valid for SGRBs as well. For the first time they concluded that the $E_{\text{peak}}-E_{\text{iso}}$ correlation for SGRBs was given by

$$\log E_{\text{iso}} = (51.42 \pm 0.15) + (1.58 \pm 0.28)\log E_{\text{peak}}^*, \tag{3.62}$$

with $r = 0.91$, $P = 1.5 \times 10^{-3}$, E_{iso} in erg s^{-1} and E_{peak}^* in units of 774.5 keV. Furthermore, the $E_{\text{peak}}-L_{\text{peak}}$ correlation for SGRBs in equation (3.58) is more significant than the $E_{\text{peak}}-E_{\text{iso}}$ correlation. For an equal value of E_{peak}, SGRBs are dimmer than LGRBs by $\simeq100$ for the $E_{\text{peak}}-E_{\text{iso}}$ correlation and $\simeq5$ for the $E_{\text{peak}}-L_{\text{peak}}$ correlation.

3.3.8 The $L_{X,p}-T_{X,p}^*$ correlation and its physical interpretation

W07 developed a phenomenological model to compute both prompt and afterglow variables contemporaneously. Both parts are nicely described by the same function:

$$f_i(t) = \begin{cases} F_i e^{\alpha_i\left(1-\frac{t}{T_i}\right)}e^{-\frac{t_i}{t}}, & t < T_i, \\ F_i\left(\dfrac{t}{T_i}\right)^{-\alpha_i} e^{-\frac{t_i}{t}}, & t \geqslant T_i. \end{cases} \tag{3.63}$$

The index i denotes p or a representing, the prompt and afterglow phases, respectively. The whole light curve, $f_{\text{tot}}(t) = f_p(t) + f_a(t)$, is characterized by two groups of four quantities each: $\{T_i, F_i, \alpha_i, t_i\}$, where α_i is the temporal power-law decay index, the time t_i is the initial rise timescale, F_i is the flux and T_i is the break time. Figure 1.8 depicts this function.

Following Dainotti *et al* (2008), employing 107 GRB light curves detected by the XRT, Qi and Lu (2010) studied the prompt phase, discovering a correlation between $L_{X,p}$ and $T_{X,p}^*$. Due to the lack of redshift measurements and solid prompt spectral parameters only 47 GRBs were employed in the final sample. Among those there were 37 GRBs with $T_{X,p}^* > 2$ s and three GRBs with $T_{X,p}^* > 100$ s.

They obtained that the correlation can have the form:

$$\log L_{X,p} = a + b \log T_{X,p}^*, \tag{3.64}$$

where $L_{X,p}$ is in erg s^{-1} and $T^*_{X,p}$ is in seconds. The D'Agostini (2005) fitting method was employed on the samples:

1. The whole dataset of 47 GRBs (see the left panel of figure 3.12).
2. 37 GRBs with $T^*_{X,p} > 2$ s (see the middle panel of figure 3.12).
3. 34 GRBs with 2 s $< T^*_{X,p} < 100$ s (see the right panel of figure 3.12).

From these fits different trends compared to equation (3.64) were found. For the first sample, the results of the fit yield a normalization $a = 50.91 \pm 0.23$ and a slope $b = -0.89 \pm 0.19$. For the other two samples, other values were recovered, $b = -1.73$ and $b = -0.74$, respectively. It was observed that the third sample gave the smallest scatter value σ_{int}, and most importantly a slope ($-0.74^{+0.20}_{-0.19}$) comparable within 1σ to that of the LT correlation (Dainotti *et al* 2008).

As claimed by Qi and Lu (2010), an indication of curvature appears in the $L_{X,p}$–$T^*_{X,p}$ correlation (see the middle panel of figure 3.12). Particularly, all GRBs which present $T^*_{X,p} < 2$ s are placed below the best-fit line, but the small number of GRBs used in their study does not allow for any definitive statements on the reliability of this effect. Indeed, it may be due to selection effects from outliers. In the case the effect is real, $T^*_{X,p}$ could be used to pursue a new classification of GRBs into LGRBs and SGRBs, as already suggested by O'Brien and Willingale (2007). Both T_{90} and $T_{X,p}$ evaluate the GRB duration: T_{90} is based on the energetics of the GRB, while $T_{X,p}$ is based on the GRB time evolution of the light curves. Nevertheless, substituting $T_{X,p}$ with T_{90} does not allow for the recovery of the correlation. For an investigation of an updated dataset and comparison of $T_{X,p}$ with T_{45} see Dainotti *et al* (2011b).

To physically interpret this correlation, it is necessary recall that both the existence of a few GRBs with a high value of $T^*_{X,p}$ and several physical emission processes can be the reason behind this curvature in the $L_{X,p}$–$T^*_{X,p}$ correlation in Qi and Lu (2010). However, this issue cannot be solved due to the small number of GRBs in the given dataset and the existence of possible outliers in the sample. To better confirm this correlation and its interpretation, further study is essential.

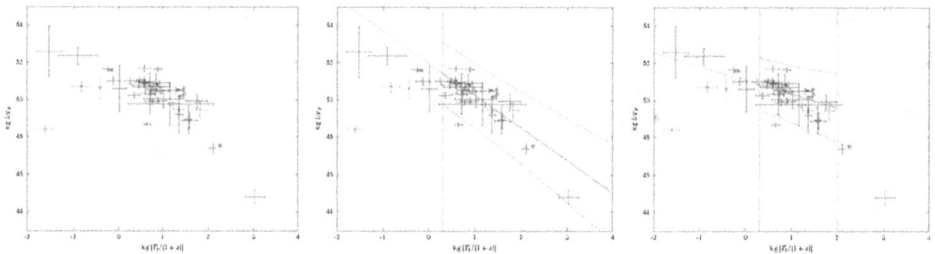

Figure 3.12. Left panel: the $L_{X,p}$–log $T^*_{X,p}$ correlation for the set of 47 GRBs (i.e. $T_{90} < 2$ s). Middle panel: best-fit of the $L_{X,p}$–log $T^*_{X,p}$ correlation for only GRBs with $T^*_{X,p} > 2$ s. Right panel: best-fit of the $L_{X,p}$–log $T^*_{X,p}$ correlation for the 34 GRBs with 2 s $< T^*_{X,p} < 100$ s. (Reproduced with permission from Qi and Lu (2010). Copyright 2010 The American Astronomical Society.)

3.3.9 The $L_{X,f}$–$T_{X,f}$ correlation and its physical interpretation

As claimed by Nousek *et al* (2006), a FRED phase following the prompt emission is detected in several GRBs. Due to the fact that this FRED is smoothly prolonged after the prompt phase, as concluded by O'Brien *et al* (2006), it can be considered as the prompt phase's tail. Among the several models (see Zhang *et al* 2007b) trying to investigate the FRED, is included the high-latitude emission (HLE) model. The HLE scenario asserts that given the additional path length due to the curvature of the emitting region, the photons reach the observer from angles which appear to be wider with respect to the line of sight. This happens when the prompt phase radiation from a spherical shell switches off at a given radius. In such a situation the Doppler factor of these photons is small.

Willingale *et al* (2010) improved the procedure developed by W07 to investigate the pulses in the prompt phase and the late X-ray flares detected by *Swift*/BAT +XRT. The pulse profile is fitted by this function:

$$
P = \left\{ \left[\min\left(\frac{T - T_{ej}}{T_{X,f}}, 1 \right)^{\alpha+2} - \left(\frac{T_{X,f} - T_{rise}}{T_{X,f}} \right)^{\alpha+2} \right] \right.
$$
$$
\left. \times \left[1 - \left(\frac{T_{X,f} - T_{rise}}{T_{X,f}} \right)^{\alpha+2} \right]^{-1} \right\} \left(\frac{T - T_{ej}}{T_{X,f}} \right)^{-1},
\tag{3.65}
$$

where $T_0 = T_{X,f} - T_{rise}$ is the time of the arrival of the first photon radiated from the shell, and T_{rise} the rise time of the pulse. The emission is supposed to originate from an ultra-relativistic thin shell expanding on limited values of radii along the line of sight, computed in the observer frame since the ejection time, T_{ej}. In this way, the rise of the pulse can be shaped using α, T_{rise} and $T_{X,f}$.

The rise and decay of the pulse is due to the union of the pulse profile $P(t, T_{X,f}, T_{rise})$ and $B(x)$, roughly represented by the Band function:

$$
B(x) = B_{norm} \times \begin{cases} x^{(\alpha-1)} e^{-x}, & x \leqslant \alpha - \beta \\ x^{(\beta-1)} (\alpha - \beta)^{(\alpha-\beta)} e^{-(\alpha-\beta)}, & x > \alpha - \beta \end{cases},
\tag{3.66}
$$

where $x = (E/E_{X,f})[(T - T_{ej})/T_{X,f}]^{-1}$, $E_{X,f}$ is the spectral break energy and B_{norm} the normalization.

Considering a sample of 12 *Swift* GRBs detected in the BAT and XRT energy ranges, Willingale *et al* (2010) discovered that $L_{X,f}$ is anti-correlated with $T_{X,f}^*$ through the equation:

$$
\log L_{X,f}^* \sim -(2.0 \pm 0.2) \log T_{X,f}^*.
\tag{3.67}
$$

This correlation shows that bright pulses take place just after ejection, while dim pulses take place at later time (see the left panel of figure 3.13). In Willingale *et al* (2010) a correlation between $L_{X,f}$ and E_{peak} was also claimed, as shown in the right panel of figure 3.13, in accordance with the already known Yonetoku correlation

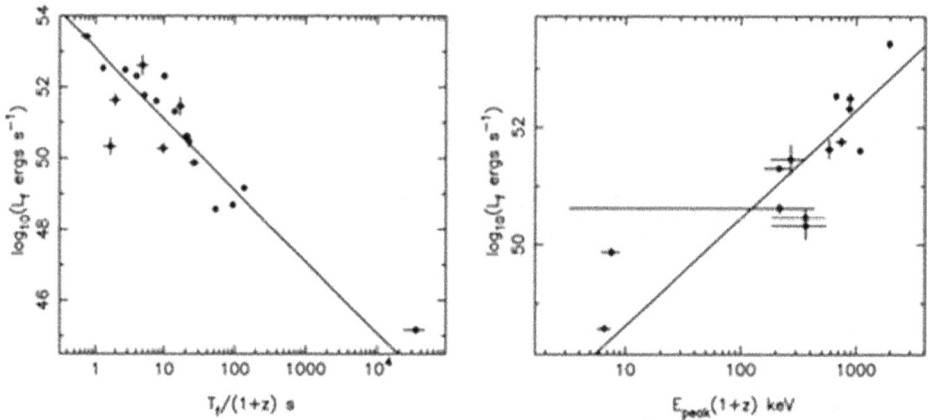

Figure 3.13. Left panel: the $L_{X,f}$–$T^*_{X,f}$ distribution. Right panel: $L_{X,f}$–E_{peak} distribution. (Reproduced from Willingale *et al* (2010). By permission of Oxford University Press on behalf of the Royal Astronomical Society.)

between L_{peak} and E_{peak} of the spectrum during time T_{90} (Yonetoku *et al* 2004, Tsutsui *et al* 2013), see also section 3.3.6.

From the sample analysed by Willingale *et al* (2010), 49 pulses were adequately fitted with the FRED function indicating the HLE model as the correct one to describe the GRB prompt phase, except for some hard peak pulses. However, an anomaly is represented by the hard pulse in GRB 061121 which demands a spectral index $\beta_S = 2.4$, which is different from the value assumed for synchrotron emission, i.e. $\beta_S = 1$.

Examples of similar correlations were given by Lee *et al* (2000) and Quilligan *et al* (2002), who analysed the width of a pulse instead of $T_{X,f}$. Later, other authors (Littlejohns *et al* 2013, Bošnjak and Daigne 2014, Evans *et al* 2014, Hakkila and Preece 2014, Laskar *et al* 2014, Littlejohns and Butler 2014, Roychoudhury *et al* 2014, Ceccobello and Kumar 2015, Kazanas *et al* 2015, Laskar *et al* 2015, Peng *et al* 2015) employed the pulse profile by Willingale *et al* (2010) to investigate the prompt radiation features of the pulses.

Regarding the physical interpretation of this correlation, in Willingale *et al* (2010) the emission is assumed to be generated from a thin shell, as previously seen in this section. Furthermore, Genet and Granot (2009) showed that the HLE is the right model to explain the decaying phase after the peak along the shell, which is slowed down and changed by the different Doppler factor responsible for the curvature of the surface (Ryde and Petrosian 2002, Dermer 2007).

References

Abbott B P *et al* 2016 Observation of gravitational waves from a binary black hole merger *Phys. Rev. Lett.* **116** 061102

Abbott B P *et al* 2017a Gravitational waves and gamma-rays from a binary neutron star merger: GW170817 and GRB 170817A *Astrophys. J.* **848** L13

Abbott B P *et al* 2017b A gravitational-wave standard siren measurement of the Hubble constant *Nature* **551** 85–8

Abdo A A *et al* 2009a *Fermi* observations of GRB 090902B: a distinct spectral component in the prompt and delayed emission *Astrophys. J.* **706** L138–44

Ackermann M *et al* 2010 *Fermi* observations of GRB 090510: a short-hard gamma-ray burst with an additional, hard power-law component from 10 keV to GeV energies *Astrophys. J.* **716** 1178–90

Ackermann M *et al* 2011 Detection of a spectral break in the extra hard component of GRB 090926A *Astrophys. J.* **729** 114

Ackermann M *et al* 2013 The first *Fermi*-LAT catalog of sources above 10 GeV *Astrophys. J. Suppl. Ser.* **209** 34

Ade P A R *et al* (Planck Collaboration) 2016 Planck 2015 results. 8. Cosmological parameters *Astron. Astrophys.* **594** A13

Amati L 2006 The $E_{p,i}E_{iso}$ correlation in gamma-ray bursts: updated observational status, re-analysis and main implications *Mon. Not. R. Astron. Soc.* **372** 233–45

Amati L 2012 Cosmology with the $E_{p,i}-E_{iso}$ correlation of gamma-ray bursts *Int. J. Mod. Phys. Conf. Ser.* **12** 19–27

Amati L and Della Valle M 2013 Measuring cosmological parameters with gamma ray bursts *Int. J. Mod. Phys.* D **22** 1330028

Amati L *et al* 2002 Intrinsic spectra and energetics of *BeppoSAX* gamma-ray bursts with known redshifts *Astron. Astrophys.* **390** 81–9

Amati L *et al* 2003 The prompt and afterglow emission of GRB 001109 measured by *BeppoSAX* *AIP Conf. Ser.* **662** 387–9

Amati L, Guidorzi C, Frontera F, Della Valle M, Finelli F, Landi R and Montanari E 2008 Measuring the cosmological parameters with the $E_{p,i}-E_{iso}$ correlation of gamma-ray bursts *Mon. Not. R. Astron. Soc.* **391** 577–84

Amati L, Frontera F and Guidorzi C 2009 Extremely energetic *Fermi* gamma-ray bursts obey spectral energy correlations *Astron. Astrophys.* **508** 173–80

Band D *et al* 1993 BATSE observations of gamma-ray burst spectra. I—Spectral diversity *Astrophys. J.* **413** 281–92

Barkana R and Loeb A 2004 Gamma-ray bursts versus quasars: Lyα signatures of reionization versus cosmological infall *Astrophys. J.* **601** 64–77

Basak R and Rao A R 2012 Correlation between the isotropic energy and the peak energy at zero fluence for the individual pulses of gamma-ray bursts: toward a universal physical correlation for the prompt emission *Astrophys. J.* **749** 132

Basak R and Rao A R 2013 Pulse-wise Amati correlation in *Fermi* gamma-ray bursts *Mon. Not. R. Astron. Soc.* **436** 3082–8

Beloborodov A M 2010 Collisional mechanism for gamma-ray burst emission *Mon. Not. R. Astron. Soc.* **407** 1033–47

Bevington P R and Robinson D K 2003 *Data Reduction and Error Analysis for the Physical Sciences* (New York: McGraw-Hill)

Bloom J S, Frail D A and Sari R 2001 The prompt energy release of gamma-ray bursts using a cosmological *k*-correction *Astronom. J.* **121** 2879–88

Borgonovo L and Ryde F 2001 On the hardness–intensity correlation in gamma-ray burst pulses *Astrophys. J.* **548** 770–86

Bošnjak Ž and Daigne F 2014 Spectral evolution in gamma-ray bursts: predictions of the internal shock model and comparison to observations *Astron. Astrophys.* **568** A45

Bromm V and Loeb A 2006 High-redshift gamma-ray bursts from population III progenitors *Astrophys. J.* **642** 382–8

Ceccobello C and Kumar P 2015 Inverse-Compton drag on a highly magnetized GRB jet in stellar envelope *Mon. Not. R. Astron. Soc.* **449** 2566–75

Collazzi A C, Schaefer B E and Moree J A 2011 The total errors in measuring E_{peak} for gamma-ray bursts *Astrophys. J.* **729** 89

Cucchiara A *et al* 2011 A photometric redshift of $z \sim 9.4$ for GRB 090429B *Astrophys. J.* **736** 7

D'Agostini G 2005 Fits, and especially linear fits, with errors on both axes, extra variance of the data points and other complications, arXiv:physics/0511182

Daigne F and Mochkovitch R 2003 The physics of pulses in gamma-ray bursts: emission processes, temporal profiles and time-lags *Mon. Not. R. Astron. Soc.* **342** 587–92

Dainotti M G and Del Vecchio R 2017 Gamma ray burst afterglow and prompt–afterglow relations: an overview *New Astron. Rev.* **77** 23–61

Dainotti M G, Cardone V F and Capozziello S 2008 A time–luminosity correlation for γ-ray bursts in the X-rays *Mon. Not. R. Astron. Soc.* **391** L79–83

Dainotti M G, Willingale R, Capozziello S, Fabrizio Cardone V and Ostrowski M 2010 Discovery of a tight correlation for gamma-ray burst afterglows with 'canonical' light curves *Astrophys. J.* **722** L215–9

Dainotti M G, Fabrizio Cardone V, Capozziello S, Ostrowski M and Willingale R 2011a Study of possible systematics in the L^*_X–T^*_a correlation of gamma-ray bursts *Astrophys. J.* **730** 135

Dainotti M G, Ostrowski M and Willingale R 2011b Towards a standard gamma-ray burst: tight correlations between the prompt and the afterglow plateau phase emission *Mon. Not. R. Astron. Soc.* **418** 2202–6

Dainotti M G, Cardone V F, Piedipalumbo E and Capozziello S 2013a Slope evolution of GRB correlations and cosmology *Mon. Not. R. Astron. Soc.* **436** 82–8

Dainotti M G, Del Vecchio R and Tarnopolski M 2018 Gamma-ray burst prompt correlations *Adv. Astron.* **2018** 4969503

de Souza R S, Yoshida N and Ioka K 2011 Populations III.1 and III.2 gamma-ray bursts: constraints on the event rate for future radio and X-ray surveys *Astron. Astrophys.* **533** A32

Dermer C D 2007 Rapid X-ray declines and plateaus in *Swift* GRB light curves explained by a highly radiative blast wave *Astrophys. J.* **664** 384–96

Evans P A *et al* 2014 GRB 130925A: an ultralong gamma ray burst with a dust-echo afterglow, and implications for the origin of the ultralong GRBs *Mon. Not. R. Astron. Soc.* **444** 250–67

Fan Y-Z 2010 The spectrum of γ-ray burst: a clue *Mon. Not. R. Astron. Soc.* **403** 483–90

Fenimore E E and Ramirez-Ruiz E 2000 Redshifts for 220 BATSE gamma-ray bursts determined by variability and the cosmological consequences, arXiv:astro-ph/0004176

Fenimore E E, Madras C D and Nayakshin S 1996 Expanding relativistic shells and gamma-ray burst temporal structure *Astrophys. J.* **473** 998

Fishman G J *et al* 1994 The first BATSE gamma-ray burst catalog *Astrophys. J. Suppl. Ser.* **92** 229–83

Freedman D L and Waxman E 2001 On the energy of gamma-ray bursts *Astrophys. J.* **547** 922–8

Frontera F *et al* 2000 Prompt and delayed emission properties of gamma-ray bursts observed with *BeppoSAX Astrophys. J. Suppl. Ser.* **127** 59–78

Frontera F, Amati L, Guidorzi C, Landi R and in't Zand J 2012 Broadband time-resolved $E_{p,i}$-L_{iso} correlelatiion in gamma-ray bursts *Astrophys. J.* **754** 138

Frontera F, Amati L, Farinelli R, Dichiara S, Guidorzi C, Landi R and Titarchuk L 2016 Possible physical explanation of the intrinsic $E_{p,i}$-'intensity' correlation commonly used to 'standardize' GRBs *Int. J. Mod. Phys.* D **25** 1630014

Gehrels N *et al* 2006 A new γ-ray burst classification scheme from GRB060614 *Nature* **444** 1044–6

Genet F and Granot J 2009 Realistic analytic model for the prompt and high-latitude emission in GRBs *Mon. Not. R. Astron. Soc.* **399** 1328–46

Ghirlanda G, Celotti A and Ghisellini G 2003 Extremely hard GRB spectra prune down the forest of emission models *Astron. Astrophys.* **406** 879–92

Ghirlanda G, Ghisellini G and Celotti A 2004a The spectra of short gamma-ray bursts *Astron. Astrophys.* **422** L55–8

Ghirlanda G, Ghisellini G and Lazzati D 2004b The collimation-corrected gamma-ray burst energies correlate with the peak energy of their νF_ν spectrum *Astrophys. J.* **616** 331–8

Ghirlanda G, Ghisellini G and Firmani C 2005a Probing the existence of the E_{peak}–E_{iso} correlation in long gamma ray bursts *Mon. Not. R. Astron. Soc.* **361** L10–4

Ghirlanda G, Ghisellini G, Firmani C, Celotti A and Bosnjak Z 2005b The peak luminosity–peak energy correlation in gamma-ray bursts *Mon. Not. R. Astron. Soc.* **360** L45–9

Ghirlanda G, Nava L, Ghisellini G and Firmani C 2007 Confirming the γ-ray burst spectral-energy correlations in the era of multiple time breaks *Astron. Astrophys.* **466** 127–36

Ghirlanda G, Nava L, Ghisellini G, Firmani C and Cabrera J I 2008 The E_{peak}–E_{iso} plane of long gamma-ray bursts and selection effects *Mon. Not. R. Astron. Soc.* **387** 319–30

Ghirlanda G, Nava L and Ghisellini G 2010 Spectral–luminosity relation within individual *Fermi* gamma rays bursts *Astron. Astrophys.* **511** A43

Ghirlanda G, Ghisellini G, Nava L and Burlon D 2011 Spectral evolution of *Fermi*/GBM short gamma-ray bursts *Mon. Not. R. Astron. Soc.* **410** L47–51

Ghirlanda G, Ghisellini G, Salvaterra R, Nava L, Burlon D, Tagliaferri G, Campana S, D'Avanzo P and Melandri A 2013 The faster the narrower: characteristic bulk velocities and jet opening angles of gamma-ray bursts *Mon. Not. R. Astron. Soc.* **428** 1410–23

Goldstein A, Preece R D and Briggs M S 2010 A new discriminator for gamma-ray burst classification: the E_{peak}–fluence energy ratio *Astrophys. J.* **721** 1329–32

Golenetskii S V, Mazets E P, Aptekar R L and Ilinskii V N 1983 Correlation between luminosity and temperature in gamma-ray burst sources *Nature* **306** 451–3

Granot J, Piran T and Sari R 1999 Images and spectra from the interior of a relativistic fireball *Astrophys. J.* **513** 679–89

Guidorzi C, Frontera F, Montanari E, Rossi F, Amati L, Gomboc A, Hurley K and Mundell C G 2005 The gamma-ray burst variability–peak luminosity correlation: new results *Mon. Not. R. Astron. Soc.* **363** 315–25

Guidorzi C, Frontera F, Montanari E, Rossi F, Amati L, Gomboc A and Mundell C G 2006 The slope of the gamma-ray burst variability/peak luminosity correlation *Mon. Not. R. Astron. Soc.* **371** 843–51

Guiriec S *et al* 2010 Time-resolved spectroscopy of the three brightest and hardest short gamma-ray bursts observed with the *Fermi* gamma-ray burst monitor *Astrophys. J.* **725** 225–41

Guiriec S *et al* 2011 Detection of a thermal spectral component in the prompt emission of GRB 100724B *Astrophys. J.* **727** L33

Guiriec S *et al* 2013 Evidence for a photospheric component in the prompt emission of the short GRB 120323A and its effects on the GRB hardness–luminosity relation *Astrophys. J.* **770** 32

Guiriec S *et al* 2015a Toward a better understanding of the GRB phenomenon: a new model for GRB prompt emission and its effects on the new $L_i^{NT}-E_{\mathrm{peak},i}^{\mathrm{rest},NT}$ relation *Astrophys. J.* **807** 148

Guiriec S, Mochkovitch R, Piran T, Daigne F, Kouveliotou C, Racusin J, Gehrels N and McEnery J 2015b GRB 131014A: a laboratory for studying the thermal-like and non-thermal emissions in gamma-ray bursts, and the new $L^{n\mathrm{Th}}_i-E^{n\mathrm{Th},\mathrm{rest}}_{\mathrm{peak},i}$ relation *Astrophys. J.* **814** 10

Guiriec S, Gonzalez M M, Sacahui J R, Kouveliotou C, Gehrels N and McEnery J 2016 *CGRO/BATSE* data support the new paradigm for GRB prompt emission and the new $L_i^{n\mathrm{Th}}-E_{\mathrm{peak},i}^{n\mathrm{Th},\mathrm{rest}}$ relation *Astrophys. J.* **819** 79

Hakkila J and Preece R D 2014 Gamma-ray burst pulse shapes: evidence for embedded shock signatures? *Astrophys. J.* **783** 88

Hakkila J, Giblin T W, Norris J P, Fragile P C and Bonnell J T 2008 Correlations between lag, luminosity, and duration in gamma-ray burst pulses *Astrophys. J.* **677** L81

Hascoët R, Daigne F and Mochkovitch R 2013 Prompt thermal emission in gamma-ray bursts *Astron. Astrophys.* **551** A124

Heussaff V, Atteia J-L and Zolnierowski Y 2013 The $E_{\mathrm{peak}}-E_{\mathrm{iso}}$ relation revisited with *Fermi* GRBs. Resolving a long-standing debate? *Astron. Astrophys.* **557** A100

Holz D E and Hughes S A 2005 Using gravitational-wave standard sirens *Astrophys. J.* **629** 15–22

Inoue S, Omukai K and Ciardi B 2007 The radio to infrared emission of very high redshift gamma-ray bursts: probing early star formation through molecular and atomic absorption lines *Mon. Not. R. Astron. Soc.* **380** 1715–28

Ioka K and Mészáros P 2005 Radio afterglows of gamma-ray bursts and hypernovae at high redshift and their potential for 21 centimeter absorption studies *Astrophys. J.* **619** 684–96

Ioka K and Nakamura T 2001 Peak luminosity–spectral lag relation caused by the viewing angle of the collimated gamma-ray bursts *Astrophys. J.* **554** L163–7

Ito H, Nagataki S, Ono M, Lee S-H, Mao J, Yamada S, Pe'er A, Mizuta A and Harikae S 2013 Photospheric emission from stratified jets *Astrophys. J.* **777** 62

Kazanas D, Racusin J L, Sultana J and Mastichiadis A 2015 The statistics of the prompt-to-afterglow GRB flux ratios and the supercritical pile GRB model, arXiv:1501.01221

Kendall M G 1938 A new measure of rank correlation *Biometrika* **30** 81–93

Kistler M D, Yüksel H, Beacom J F, Hopkins A M and Wyithe J S B 2009 The star formation rate in the reionization era as indicated by gamma-ray bursts *Astrophys. J.* **705** L104–8

Kocevski D and Liang E 2003 The connection between spectral evolution and gamma-ray burst lag *Astrophys. J.* **594** 385–9

Kouveliotou C, Meegan C A, Fishman G J, Bhat N P, Briggs M S, Koshut T M, Paciesas W S and Pendleton G N 1993 Identification of two classes of gamma-ray bursts *Astrophys. J.* **413** L101–4

Lamb D Q, Donaghy T Q and Graziani C 2004 A unified jet model of X-ray flashes and γ-ray bursts *New Astron. Rev.* **48** 459–64

Laskar T *et al* 2014 GRB 120521C at $z \sim 6$ and the properties of high-redshift γ-ray bursts *Astrophys. J.* **781** 1

Laskar T, Berger E, Margutti R, Perley D, Zauderer B A, Sari R and Fong W-F 2015 Energy injection in gamma-ray burst afterglows *Astrophys. J.* **814** 1

Lee A, Bloom E D and Petrosian V 2000 Properties of gamma-ray burst time profiles using pulse decomposition analysis *Astrophys. J. Suppl. Ser.* **131** 1–19

Lee T T and Petrosian V 1996 Distributions of peak flux and duration for gamma-ray bursts *Astrophys. J.* **470** 479

Lei W H, Wang D X, Zhang L, Gan Z M, Zou Y C and Xie Y 2009 Magnetically torqued neutrino-dominated accretion flows for gamma-ray bursts *Astrophys. J.* **700** 1970–6

Li L-X and Paczyński B 2006 Improved correlation between the variability and peak luminosity of gamma-ray bursts *Mon. Not. R. Astron. Soc.* **366** 219–26

Liang E and Kargatis V 1996 Dependence of the spectral evolution of γ-ray bursts on their photon fluence *Nature* **381** 49–51

Liang E and Zhang B 2005 Model-independent multivariable gamma-ray burst luminosity indicator and its possible cosmological implications *Astrophys. J.* **633** 611–23

Liang E W, Dai Z G and Wu X F 2004 The luminosity–E_p relation within gamma-ray bursts and the implications for fireball models *Astrophys. J.* **606** L29–32

Liang E-W, Yi S-X, Zhang J, Lü H-J, Zhang B-B and Zhang B 2010 Constraining gamma-ray burst initial Lorentz factor with the afterglow onset feature and discovery of a tight Γ_0–$E_{\mathrm{gamma,iso}}$ correlation *Astrophys. J.* **725** 2209–24

Lin H-N, Li X, Wang S and Chang Z 2015 Are long gamma-ray bursts standard candles? *Mon. Not. R. Astron. Soc.* **453** 128–32

Lin H-N, Li X and Chang Z 2016b Effect of GRB spectra on the empirical luminosity correlations and the GRB Hubble diagram *Mon. Not. R. Astron. Soc.* **459** 2501

Lithwick Y and Sari R 2001 Lower limits on Lorentz factors in gamma-ray bursts *Astrophys. J.* **555** 540–5

Littlejohns O M and Butler N R 2014 Investigating signatures of cosmological time dilation in duration measures of prompt gamma-ray burst light curves *Mon. Not. R. Astron. Soc.* **444** 3948–60

Littlejohns O M, Tanvir N R, Willingale R, Evans P A, O'Brien P T and Levan A J 2013 Are gamma-ray bursts the same at high redshift and low redshift? *Mon. Not. R. Astron. Soc.* **436** 3640–55

Lloyd N M, Petrosian V and Mallozzi R S 2000a Cosmological versus intrinsic: the correlation between intensity and the peak of the $\nu F_{\frac{1}{2}}$ spectrum of gamma-ray bursts *Astrophys. J.* **534** 227–38

Lloyd N M, Petrosian V and Preece R D 2000b Synchrotron emission as the source of GRB spectra. Part II: observations *AIP Conf. Ser.* **662** 155–9

Lloyd-Ronning N M and Petrosian V 2002 Interpreting the behavior of time-resolved gamma-ray burst spectra *Astrophys. J.* **565** 182–94

Lloyd-Ronning N M and Ramirez-Ruiz E 2002 On the spectral energy dependence of gamma-ray burst variability *Astrophys. J.* **576** 101–6

Lü J, Zou Y-C, Lei W-H, Zhang B, Wu Q, Wang D-X, Liang E-W and Lü H-J 2012 Lorentz-factor-isotropic–luminosity/energy correlations of gamma-ray bursts and their interpretation *Astrophys. J.* **751** 49

Lu R and Liang E 2010 Luminosity–peak energy relation in the decay phases of gamma-ray burst pulses *Sci. China Phys. Mech. Astron.* **53** 163–70

Lu R-J, Wei J-J, Liang E-W, Zhang B-B, Lü H-J, Lü L-Z, Lei W-H and Zhang B 2012 A comprehensive analysis of *Fermi* gamma-ray burst data. II. E_p evolution patterns and implications for the observed spectrum–luminosity relations *Astrophys. J.* **756** 112

Mallozzi R S, Paciesas W S, Pendleton G N, Briggs M S, Preece R D, Meegan C A and Fishman G J 1995 The $\nu F\nu$ peak energy distributions of gamma-ray bursts observed by BATSE *Astrophys. J.* **454** 597

Margutti R, Guidorzi C, Chincarini G, Bernardini M G, Genet F, Mao J and Pasotti F 2010 Lag–luminosity relation in γ-ray burst X-ray flares: a direct link to the prompt emission *Mon. Not. R. Astron. Soc.* **406** 2149–67

Mendoza S, Hidalgo J C, Olvera D and Cabrera J I 2009 Internal shocks in relativistic jets with time-dependent sources *Mon. Not. R. Astron. Soc.* **395** 1403–8

Mészáros P 2006 Gamma-ray bursts *Rep. Prog. Phys.* **69** 2259–321

Mochkovitch R and Nava L 2015 The E_p–E_{iso} relation and the internal shock model *Astron. Astrophys.* **577** A31

Nava L, Ghisellini G, Ghirlanda G, Tavecchio F and Firmani C 2006 On the interpretation of spectral-energy correlations in long gamma-ray bursts *Astron. Astrophys.* **450** 471–81

Nava L *et al* 2012 A complete sample of bright *Swift* long gamma-ray bursts: testing the spectral-energy correlations *Mon. Not. R. Astron. Soc.* **421** 1256–64

Norris J P, Marani G F and Bonnell J T 2000 Connection between energy-dependent lags and peak luminosity in gamma-ray bursts *Astrophys. J.* **534** 248–57

Nousek J A *et al* 2006 Evidence for a canonical gamma-ray burst afterglow light curve in the *Swift* XRT data *Astrophys. J.* **642** 389–400

O'Brien P T and Willingale R 2007 Using *Swift* observations of prompt and afterglow emission to classify GRBs *R. Soc. Lond. Philos. Trans. Ser.* A **365** 1179–88

O'Brien P T *et al* 2006 The early X-ray emission from GRBs *Astrophys. J.* **647** 1213–37

Panaitescu A 2009 An external-shock origin of the relation for gamma-ray bursts *Mon. Not. R. Astron. Soc.* **393** 1010–5

Panaitescu A and Kumar P 2002 Properties of relativistic jets in gamma-ray burst afterglows *Astrophys. J.* **571** 779–89

Peng Z Y, Yin Y, Yi T F, Bao Y Y and Wu H 2015 A comprehensive comparative study of temporal properties between X-ray flares and GRB pulses *Astrophys. Space Sci.* **355** 95–103

Perlmutter S *et al* 1998 Discovery of a supernova explosion at half the age of the Universe *Nature* **391** 51

Piran T 2004 The physics of gamma-ray bursts *Rev. Mod. Phys.* **76** 1143–210

Porciani C and Madau P 2011 On the association of gamma-ray bursts with massive stars: implications for number counts and lensing statistics *Astrophys. J.* **548** 522–31

Preece R D, Briggs M S, Mallozzi R S, Pendleton G N, Paciesas W S and Band D L 2000 The BATSE gamma-ray burst spectral catalog. I. High time resolution spectroscopy of bright bursts using high energy resolution data *Astrophys. J. Suppl. Ser.* **126** 19–36

Qi S and Lu T 2010 A new luminosity relation for gamma-ray bursts and its implication *Astrophys. J.* **717** 1274–8

Qin Y-P and Chen Z-F 2013 Statistical classification of gamma-ray bursts based on the Amati relation *Mon. Not. R. Astron. Soc.* **430** 163–73

Quilligan F, McBreen B, Hanlon L, McBreen S, Hurley K J and Watson D 2002 Temporal properties of gamma ray bursts as signatures of jets from the central engine *Astron. Astrophys.* **385** 377–98

Ramirez-Ruiz E 2005 Photospheric signatures imprinted on the γ-ray burst spectra *Mon. Not. R. Astron. Soc.* **363** L61–5

Rees M J and Mészáros P 2005 Dissipative photosphere models of gamma-ray bursts and X-ray flashes *Astrophys. J.* **628** 847–52

Reichart D E and Nysewander M C 2005 GRB variability–luminosity correlation confirmed, arXiv:astro-ph/0508111

Reichart D E, Lamb D Q, Fenimore E E, Ramirez-Ruiz E, Cline T L and Hurley K 2001 A possible Cepheid-like luminosity estimator for the long gamma-ray bursts *Astrophys. J.* **552** 57–71

Riess A G *et al* 1998 Observational evidence from supernovae for an accelerating universe and a cosmological constant *Astronom. J.* **116** 1009–38

Riess A G *et al* 2016 A 2.4% determination of the local value of the Hubble constant *Astrophys. J.* **826** 56

Rizzuto D *et al* 2007 Testing the gamma-ray burst variability/peak luminosity correlation on a *Swift* homogeneous sample *Mon. Not. R. Astron. Soc.* **379** 619–28

Rodney S A *et al* 2015 Two SNe Ia at redshift ~2: improved classification and redshift determination with medium-band infrared imaging *Astronom. J.* **150** 156

Roychoudhury A, Sarkar S K and Bhadra A 2014 Spectral lag features of GRB 060814 from *Swift* BAT and Suzaku observations *Astrophys. J.* **782** 105

Ryan G, van Eerten H, MacFadyen A and Zhang B-B 2015 Gamma-ray bursts are observed off-axis *Astrophys. J.* **799** 3

Ryde F 2005 Is thermal emission in gamma-ray bursts ubiquitous? *Astrophys. J.* **625** L95–8

Ryde F and Petrosian V 2002 Gamma-ray burst spectra and light curves as signatures of a relativistically expanding plasma *Astrophys. J.* **578** 290–303

Ryde F and Svensson R 2002 On the variety of the spectral and temporal behavior of long gamma-ray burst pulses *Astrophys. J.* **566** 210–28

Sakamoto T *et al* 2004 High energy transient explorer 2 observations of the extremely soft X-ray flash XRF 020903 *Astrophys. J.* **602** 875–85

Salmonson J D 2000 On the kinematic origin of the luminosity–pulse lag relationship in gamma-ray bursts *Astrophys. J.* **544** L115–7

Salmonson J D and Galama T J 2002 Discovery of a tight correlation between pulse lag/luminosity and jet-break times: a connection between gamma-ray bursts and afterglow properties *Astrophys. J.* **569** 682–8

Salvaterra R 2015 High redshift gamma-ray bursts *J. High Energy Astrophys.* **7** 35–43

Sari R and Piran T 1999 GRB 990123: the optical flash and the fireball model *Astrophys. J.* **517** L109–112

Sari R, Piran T and Halpern J P 1999 Jets in gamma-ray bursts *Astrophys. J.* **519** L17–20

Schaefer B E 2003a Gamma-ray burst Hubble diagram to $z = 4.5$ *Astrophys. J.* **583** L67–70

Schaefer B E 2003b Explaining the gamma-ray burst E_{peak} distribution *Astrophys. J.* **583** L71–4

Schaefer B E 2004 Explaining the gamma-ray burst lag/luminosity relation *Astrophys. J.* **602** 306–11

Schaefer B E 2007 The Hubble diagram to redshift >6 from 69 gamma-ray bursts *Astrophys. J.* **660** 16–46

Schaefer B E, Deng M and Band D L 2001 Redshifts and luminosities for 112 gamma-ray bursts *Astrophys. J.* **563** L123–7

Schutz B F 1986 Determining the Hubble constant from gravitational wave observations *Nature* **323** 310

Spearman C 1904 The proof and measurement of association between two things *Am. J. Psychol.* **15** 72–101

Stern B E and Svensson R 1996 Evidence for 'chain reaction' in the time profiles of gamma-ray bursts *Astrophys. J.* **469** L109

Stratta G, Guetta D, D'Elia V, Perri M, Covino S and Stella L 2009 Evidence for an anticorrelation between the duration of the shallow decay phase of GRB X-ray afterglows and redshift *Astron. Astrophys.* **494** L9–12

Sultana J, Kazanas D and Fukumura K 2012 Luminosity correlations for gamma-ray bursts and implications for their prompt and afterglow emission mechanisms *Astrophys. J.* **758** 32

Titarchuk L, Farinelli R, Frontera F and Amati L 2012 An upscattering spectral formation model for the prompt emission of gamma-ray bursts *Astrophys. J.* **752** 116

Totani T 1997 Cosmological gamma-ray bursts and evolution of galaxies *Astrophys. J.* **486** L71–4

Tsutsui R, Nakamura T, Yonetoku D, Murakami T, Tanabe S and Kodama Y 2008 Redshift dependent lag–luminosity relation in 565 BASTE gamma ray bursts *AIP Conf. Ser.* **1000** 28–31

Tsutsui R, Nakamura T, Yonetoku D, Murakami T, Kodama Y and Takahashi K 2009a Cosmological constraints from calibrated Yonetoku and Amati relation suggest fundamental plane of gamma-ray bursts *J. Cosmol. Astropart. Phys.* **8** 015

Tsutsui R, Nakamura T, Yonetoku D, Murakami T and Takahashi K 2010 Intrisic dispersion of correlations among E_p, L_p, and E_{iso} of gamma ray bursts depends on the quality of data set, arXiv:1012.3009

Tsutsui R, Yonetoku D, Nakamura T, Takahashi K and Morihara Y 2013 Possible existence of the E_p–L_p and E_p–E_{iso} correlations for short gamma-ray bursts with a factor 5–100 dimmer than those for long gamma-ray bursts *Mon. Not. R. Astron. Soc.* **431** 1398–404

Uhm Z L and Zhang B 2016 Toward an understanding of GRB prompt emission mechanism. I. The origin of spectral lags *Astrophys. J.* **825** 97

Ukwatta T N, Stamatikos M, Dhuga K S, Sakamoto T, Barthelmy S D, Eskandarian A, Gehrels N, Maximon L C, Norris J P and Parke W C 2010 Spectral lags and the lag–luminosity relation: an investigation with *Swift* BAT gamma-ray bursts *Astrophys. J.* **711** 1073–86

Ukwatta T N *et al* 2012 The lag–luminosity relation in the GRB source frame: an investigation with *Swift* BAT bursts *Mon. Not. R. Astron. Soc.* **419** 614–23

Vurm I and Beloborodov A M 2016 Radiative transfer models for gamma-ray bursts *Astrophys. J.* **831** 175

Wang F Y, Dai Z G and Liang E W 2015 Gamma-ray burst cosmology *New Astron. Rev.* **67** 1–17

Waxman E 1997 γ-ray burst afterglow: confirming the cosmological fireball model *Astrophys. J.* **489** L33–6

Weinberg D H, Mortonson M J, Eisenstein D J, Hirata C, Riess A G and Rozo E 2013 Observational probes of cosmic acceleration *Phys. Rep.* **530** 87–255

Wijers R A M 1999 Physical parameters of GRB 970508 and GRB 971214 from their afterglow synchrotron emission *Astrophys. J.* **523** 177–86

Willingale R *et al* 2007 Testing the standard fireball model of gamma-ray bursts using late X-ray afterglows measured by *Swift Astrophys. J.* **662** 1093–110

Willingale R, Genet F, Granot J and O'Brien P T 2010 The spectral–temporal properties of the prompt pulses and rapid decay phase of gamma-ray bursts *Mon. Not. R. Astron. Soc.* **403** 1296–316

Xiao L and Schaefer B E 2009 Estimating redshifts for long gamma-ray bursts *Astrophys. J.* **707** 387–403

Yi T-F, Xie G-Z and Zhang F-W 2008 A close correlation between the spectral lags and redshifts of gamma-ray bursts *Chinese J. Astron. Astrophys.* **8** 81–6

Yonetoku D, Murakami T, Nakamura T, Yamazaki R, Inoue A K and Ioka K 2004 Gamma-ray burst formation rate inferred from the spectral peak energy–peak luminosity relation *Astrophys. J.* **609** 935–51

Yonetoku D, Murakami T, Tsutsui R, Nakamura T, Morihara Y and Takahashi K 2010 Possible origins of dispersion of the peak energy–brightness correlations of gamma-ray bursts *Publ. Astron. Soc. Jpn.* **62** 1495

Yonetoku D, Nakamura T, Sawano T, Takahashi K and Toyanago A 2014 Short gamma-ray burst formation rate from BATSE data using E_p –L_p correlation and the minimum gravitational-wave event rate of a coalescing compact binary *Astrophys. J.* **789** 65

Zhang B and Mészáros P 2002 An analysis of gamma-ray burst spectral break models *Astrophys. J.* **581** 1236–47

Zhang B and Pe'er A 2009 Evidence of an initially magnetically dominated outflow in GRB 080916C *Astrophys. J.* **700** L65–8

Zhang B and Yan H 2011 The internal-collision-induced magnetic reconnection and turbulence (ICMART) model of gamma-ray bursts *Astrophys. J.* **726** 90

Zhang B *et al* 2009 Discerning the physical origins of cosmological gamma-ray bursts based on multiple observational criteria: the cases of $z = 6.7$ GRB 080913, $z = 8.2$ GRB 090423, and some short/hard GRBs *Astrophys. J.* **703** 1696–724

Zhang B-B, Liang E-W and Zhang B 2007b A comprehensive analysis of *Swift* XRT data. I. Apparent spectral evolution of gamma-ray burst X-ray tails *Astrophys. J.* **666** 1002–11

Zhang Z, Xie G Z, Deng J G and Jin W 2006 Revisiting the characteristics of the spectral lags in short gamma-ray bursts *Mon. Not. R. Astron. Soc.* **373** 729–32

Zou Y-C and Piran T 2010 Lorentz factor constraint from the very early external shock of the gamma-ray burst ejecta *Mon. Not. R. Astron. Soc.* **402** 1854–62

Chapter 4

Selection effects on prompt correlations

4.1 Introduction to selection effects

Selection effects are distortions or biases that usually occur when the observational sample is not representative of the 'true' underlying population. These kinds of biases usually affect GRB relations. Efron and Petrosian (1992), Dainotti *et al* (2013, 2015a) and Petrosian *et al* (2013) emphasized that when dealing with a multivariate dataset, it is imperative to determine first the true relations among the variables, not those introduced by the observational selection effects, before obtaining the individual distributions of the variables themselves. In other words, it is important to focus on the intrinsic correlations between the parameters, not on the observed ones, because the latter can be just the result of selection bias due to instrumental thresholds. How lack of knowledge about the efficiency function influences the parameters of the correlations has already been discussed for both the prompt (Butler *et al* 2009) and afterglow phases (Dainotti *et al* 2015b).

This kind of study is necessary to claim the existence of intrinsic relations. A relation can be called intrinsic only if it is carefully tested and corrected for these biases.

The selection effects present in the relations discussed above are mostly due to the dependence of the parameters on the redshift, such as in the case of the time and the luminosity evolution, or due to the threshold of the detector used.

In this section, we describe several different methods to deal with selection biases.

4.2 Selection effects for peak energy

Mallozzi *et al* (1995) discussed the photon spectra used for the determination of E_{peak}. These parameters were obtained by averaging the count rate over the duration of each event. In addition, the temporal evolution of the single light curve affects the signal-to-noise ratio (S/N), i.e. a more spiky light curve will have a larger S/N than a smooth, single-peak event. However, Ford *et al* (1995) showed that this effect is not relevant for E_{peak}. Considering the most luminous GRBs, they claimed a relevant evolution of E_{peak} with time. For this reason, Mallozzi *et al* (1995) used time-averaged spectra, resulting in an average value of E_{peak} for each burst. They believed that this evolution should not have a significant impact on their results. It should be noted, however, that E_{peak} evolution for bursts with different intensities has not yet been examined. Moreover, it has been shown that the fluence, the flux and the peak energy are affected by data truncation caused by the detector threshold (Lee and Petrosian 1996, Lloyd and Petrosian 1999) and this will generate a bias against high E_{peak} bursts with small fluence or flux and an artificial positive correlation in the data. For this reason, it is important to investigate the truncation on the data before carrying out research on the correlations.

Lloyd *et al* (2000) focused on the BATSE detector to understand whether truncation effects play a role in the $E_{peak} - F_{tot}$ correlation. The threshold of any physical parameter determined by BATSE is obtained through the trigger condition. For each burst, C_{max}, C_{min}—the peak photon counts and the background in the second brightest detector—are known. Given S_{obs} (or F_{peak}), the threshold can be computed using the relation

$$\frac{F_{peak}}{F_{peak,lim}} = \frac{S_{obs}}{S_{obs,lim}} = \frac{C_{max}}{C_{min}}. \tag{4.1}$$

The condition in equation (4.1) is true if the GRB spectrum does not undergo severe spectral evolution. Lloyd *et al* (2000) adopted spectral parameters from the Band model. Given a GRB spectrum $f_{\alpha,\beta,A}(E, E_{peak}, t)$, the fluence is obtained as

$$S_{obs} = \int_0^T dt \int_{E_1}^{E_2} E f_{\alpha,\beta,A}(E, E_{peak}, t)dE, \tag{4.2}$$

where T is the burst duration. The limiting $E_{peak,lim}$, from which the BATSE instrument is still triggered, is given by the relation

$$S_{obs,lim} = \int_0^T dt \int_{E_1}^{E_2} E f_{\alpha,\beta,A}(E, E_{peak,lim}, t)dE. \tag{4.3}$$

To compute the lower ($E_{peak,min}$) and upper ($E_{peak,max}$) limits on E_{peak}, Lloyd *et al* (2000) decreased and increased, respectively, the observed value of E_{peak} until the condition in equation (4.3) was satisfied. In addition, using non-parametric techniques developed by Efron and Petrosian (1999), they showed how to correctly remove selection bias from observed correlations. This method is general for any kind of correlation. Next, similarly to what was done by Lloyd and Petrosian (1999), once $E_{peak,min}$ and $E_{peak,max}$ were determined, Lloyd *et al* (2000) showed that the intrinsic E_{peak} distribution is much broader than the observed one. Therefore, they analyzed how these biases influence the outcomes. After a careful study of the selection effects, it was claimed that an intrinsic correlation between E_{peak} and E_{iso} indeed exists. In addition, as an important constraint on physical models of GRB prompt emission, the E_{peak} distribution is broader than that inferred previously from the observed E_{peak} values of bright BATSE GRBs (Zhang and Mészáros 2002).

The E_{peak} and E_{iso} correlation, aka 'Amati relation', was actually discovered in 2002 (Amati *et al* 2002) based on the first sample of *BeppoSAX* GRBs with measured redshift, and later confirmed and extended by measurements by *HETE-2*, *Swift*, *Fermi*/GBM and KONUS/*Wind*. The fact that detectors with different sensitivities as a function of photon energy observe a similar correlation is a first-order indication that instrumental effects should not be dominant. Soon after, it was shown that the same correlation holds between E_{peak} and the peak luminosity L_{peak} (Yonetoku *et al* 2004). Moreover, it was pointed out by Ghirlanda *et al* (2004a, 2004b) that the E_{peak} and E_{iso} correlation becomes tighter and steeper (the 'Ghirlanda relation') when applying the correction for the jet opening angle. This correction, however, can be applied only for the sub-sample of GRBs from which this quantity could be estimated based on the break observed in the afterglow light curve. In addition, this method of estimating the jet opening angle is model-dependent and may be affected by uncertainties.

Later, Band and Preece (2005) showed that the Amati and the Ghirlanda relation could be converted into a similar energy ratio

$$\frac{E_{peak}^{1/\eta_i}}{S_\gamma} \propto F(z). \tag{4.4}$$

Here, η_i are the best-fit power-law indices for the respective correlations. These energy ratios can be represented as functions of redshift, $F(z)$, and their upper limits could be determined for any z. The upper limits of the energy ratio of both the Amati and Ghirlanda relations can be projected onto the peak energy–fluence plane where they become lower limits. In this way, it is possible to use GRBs without redshift measurement to test the correlations of the intrinsic peak energy E_{peak} with the radiated energy (E_{iso}, E_γ) or peak luminosity (L_{peak}), as shown in figure 4.1. By using this method the above and other authors (Goldstein *et al* 2010, Collazzi *et al* 2012) found that a significant fraction of BATSE and *Fermi* GRBs are potentially inconsistent with the E_{peak} and E_{iso} correlation.

However, several other authors (Ghirlanda *et al* 2008, Nava *et al* 2012) showed that, when properly taking into account the dispersion of the correlation and the uncertainties on spectral parameters and fluencies, only a few percent of GRBs may be outliers of the correlation. Moreover, it can be demonstrated (Dichiara *et al* 2013) that such a small fraction of outliers can be artificially created by the combination of instrumental sensitivity and energy band, namely the typical hard-to-soft spectral evolution of GRBs.

Along this line of investigation, recently, Bošnjak *et al* (2014) presented the evaluation of E_{peak} based on the updated *INTEGRAL* catalogue of GRBs observed between December 2002 and February 2012. In their spectral analysis they investigated the energy regions with the highest sensitivity to compute the spectral peak energies. To account for the possible biases in the distribution of the spectral parameters, they compared the GRBs detected by *INTEGRAL* with the ones observed by *Fermi* and BATSE within the same fluence range. A lower flux limit

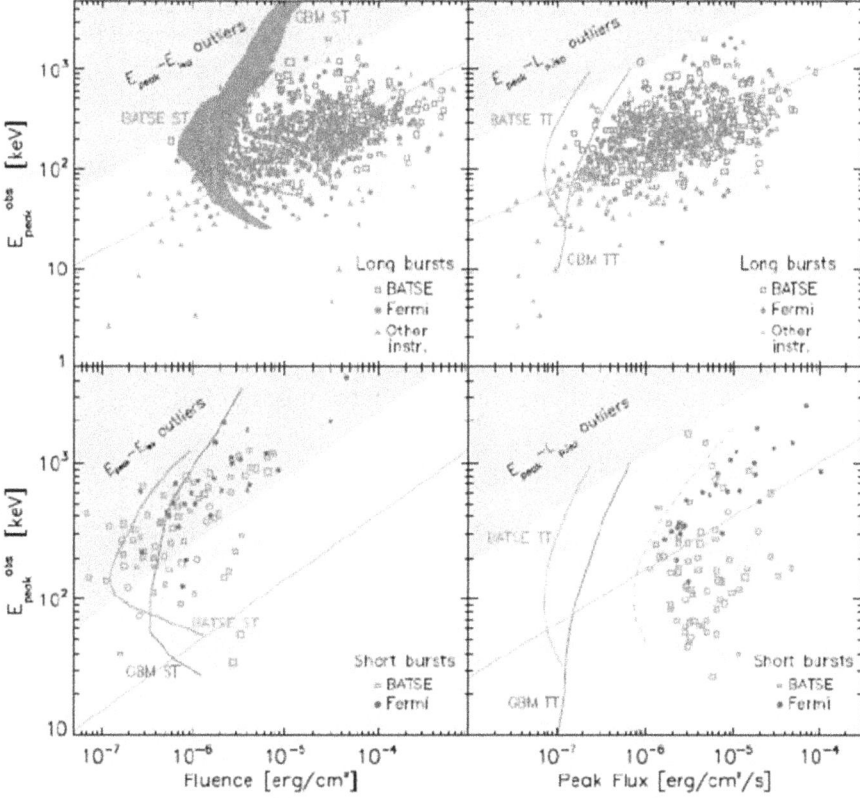

Figure 4.1. E_{peak}–fluence and E_{peak}–peak flux planes for long (upper panels) and short (bottom panels) bursts (Nava *et al* 2012). Empty squares represent BATSE bursts, filled circles represent GBM bursts and filled triangles indicate events detected by other instruments. In all panels the instrumental limits for BATSE and GMB are reported: shaded curved regions in the upper-left panel show the fluence threshold, estimated assuming burst durations of 5 and 20 s; solid curves in the bottom-left panel represent the fluence threshold for short bursts. Solid curves in the right-hand panels define the trigger threshold, identical for short and long events. The dashed curve in the bottom-right panel represents the selection criterion applied, i.e. peak flux $\geqslant 3$ photons cm^{-2} s[1]. The shaded regions in the upper-left corners of all the planes are the region identifying the outliers at more than 3σ of the $E_{peak} - E_{iso}$ (left-hand panels) and $E_{peak} - L_{peak}$ (right-hand panels) correlations for any given redshift. GRBs, without measured redshift, which fall in these regions are outliers of the corresponding rest-frame correlations ($E_{peak} - E_{iso}$ and $E_{peak} - L_{peak}$ for the left- and right-hand panels, respectively) for any assigned redshift. This means that there is no redshift, which makes them consistent with these correlations (considering their 3σ scatter). © 2018 The Astronomical Society of the Pacific. Reproduced by permission IOP Publishing. All rights reserved.

($< 8.7 \times 10^{-5}$ erg cm^{-2} in 50–300 keV energy range) was assumed because the peak fluxes from different telescopes were computed in distinct energy ranges. Then, with the proper evaluation of E_{peak}, they computed correlations between the following parameters: (i) E_{peak} and F_{tot}, (ii) E_{peak} and α, and (iii) E_0 and α.

In the case of the E_{peak}–α relation no significant correlation was found, while for the E_0–α relation there was a weak negative correlation ($\rho = -0.44$) with

$P = 1.15 \times 10^{-2}$. In the case of the $E_{peak}-F_{tot}$ relation, a weak positive correlation ($\rho = 0.50$) was found with $P = 1.88 \times 10^{-2}$. This is in agreement with the results of Kaneko *et al* (2006), who found a significant correlation between E_{peak} and F_{tot} analyzing the spectra of 350 bright BATSE GRBs with high spectral and temporal resolution. Regarding the detector-related E_{peak} uncertainties, Collazzi *et al* (2011) noticed that there is a discrepancy among the values of E_{peak} found in the literature that go beyond the 1σ uncertainty.

Finally, notwithstanding that the GRBs must be sufficiently bright to perform time-resolved spectroscopy and have known redshifts, if the process generating GRBs is independent of the brightness, then the existence of the time-resolved $E_{peak}-L_{iso}$ correlation (Ghirlanda *et al* 2010, Lu *et al* 2012, Frontera *et al* 2012), see the right panel of figure 4.1, is evidence that these $E_{peak}-$'intensity' correlations have a physical origin linked to the main emission mechanism in GRBs.

4.3 Selection effects for the isotropic energy

Regarding the selection effects related to E_{iso}, Amati *et al* (2002) found that the GRBs with measured redshift can be biased due to their paucity, and that the sensitivities and energy bands of the wide field camera (WFC) and gamma-ray burst monitor (GBM) onboard *BeppoSAX* and *Fermi*, respectively, might prefer energetic and luminous GRBs at larger redshifts, thus creating an artificial $E_{peak}-E_{iso}$ relation.

Therefore, similarly to Lloyd and Petrosian (1999), as discussed in section 4.2, Amati *et al* (2002) analyzed the $E_{peak,min}$ and $E_{peak,max}$ for which the $E_{peak}-E_{iso}$ correlation exists. If the spectral parameters are coincident with their minimum and maximum values, then it is very likely that data truncation will produce a spurious correlation.

Considering a sample of BATSE GRBs without measured redshifts, two research groups (Nakar and Piran 2005, Band and Preece 2005) claimed that around 50% (Nakar and Piran 2005) or even 80% (Band and Preece 2005) of GRBs do not obey the $E_{peak}-E_{iso}$ correlation. This is due to the fact that the selection effects may favour a burst sub-population for which the Amati relation is valid. GRBs with determined redshifts must be relatively bright and soft to be localized. In addition, it was found that selection effects were present in these GRB observations and this is the reason why only the redshifts of the GRBs obeying the relation were computed. However, other authors (Ghirlanda *et al* 2005b, Bosnjak *et al* 2005) arrived at opposite conclusions and these different results are due to considering (or not) the dispersion in the relation, and the uncertainties in E_{peak} and the fluence. Indeed, considering both these features, only some BATSE GRBs with no measured redshift may be outliers of the $E_{peak}-E_{iso}$ relation (Ghirlanda *et al* 2005a).

Later, Amati (2006) overestimated E_{peak} values because of the paucity of data below 25 keV. Indeed, if there were selection effects in the sample of *HETE-2* GRBs with known redshift, they are more plausible to occur due to detector sensitivity as a function of energy than as a function of the redshift (Amati 2006). The fact that all *Swift* GRBs with known redshift are consistent with the $E_{peak}-E_{iso}$ correlation is strong evidence against the existence of relevant selection effects. Amati (2006) justified this statement adducing the following points: (i) the *Swift*/BAT sensitivity in 15–30 keV is comparable to that of BATSE, *BeppoSAX* and *HETE-2*, and (ii) the rapid XRT localization of GRBs decreased the selection effects dependent on the redshift estimate. However, BAT gives an estimate of E_{peak} only for 15%–20% of the events. In addition, it was also claimed that the existence of sub-energetic events (such as GRB 980425 and possibly GRB 031203) with spectral characteristics are not in agreement with the obtained relation.

Ghirlanda *et al* (2008) studied the redshift evolution of the E_{peak}–E_{iso} correlation by binning the GRB sample into different redshift ranges and comparing the slopes in each bin. There is no evidence that this relation evolves with z, contrary to what was found with a smaller GRB sample. Their analysis showed, however, that the bursts detected before *Swift* are not influenced by the instrumental selection effects, while in the sample of *Swift* GRBs, the smallest fluence for which it is allowed to compute E_{peak} suffers from truncation effects in 27 out of 76 events.

Amati *et al* (2008) analyzed the scatter of the E_{peak}–E_{iso} relation at high energies and pointed out that it is not influenced by truncation effects because its normalization, computed assuming GRBs with precise E_{peak} from *Fermi*/GBM, is in agreement with those calculated from other satellites (e.g. *BeppoSAX*, *Swift*, KONUS/*Wind*). It was also checked whether E_{iso} in the 1 keV–10 MeV band can affect the E_{peak}–E_{iso} relation, but its scatter does not seem to vary. Finally, it was also pointed out that: (i) the distribution of the new sample of 95 LGRBs is consistent with previous results, (ii) in the E_{peak}–E_{iso} plane the scatter is smaller than in the E_{peak}–E_{iso} plane, but if the redshift is randomly distributed then the distributions are similar, and (iii) all LGRBs with measured redshift (except GRB 980425) detected with *Fermi*/GBM, *BeppoSAX*, *HETE-2* and *Swift*, obey the E_{peak}–E_{iso} relation (Amati *et al* 2008). An exhaustive analysis of instrumental and selection effects for the E_{peak}–E_{iso} correlation is underway and will be reported elsewhere.

Another example of an analysis of selection effects for E_{iso} was given by Butler *et al* (2009). They studied the influence of the detector threshold on the E_{peak}–E_{iso} relation, considering a set of 218 *Swift* GRBs and 56 *HETE-2* GRBs. Due to the different sensitivities of the *Swift* and *HETE-2* instruments, in the *Swift* survey more GRBs were detected. In other words, there is a deficit of data in samples observed in pre-*Swift* missions, and this possibly biases the correlations. Butler *et al* (2009) tested the reliability of a generic method for dealing with data truncation in the correlations, and afterwards they employed it to datasets obtained by *Swift* and pre-*Swift* satellites. However, *Swift* data do not rigorously satisfy the independence from redshift if there are only bright GRBs, as instead occurred for the pre-*Swift* E_{peak}–E_{iso} relation.

Later, Collazzi *et al* (2012) argued that the Amati relation may be an artifact of, or at least significantly biased by, a combination of selection effects due to detector sensitivity and energy thresholds. It was found that GRBs following the Amati relation are distributed above a limiting line. Even if bursts with spectroscopic redshifts are consistent with Amati's limit, it is not true for bursts with spectroscopic redshift measured by BATSE and *Swift*. In the case in which selection effects are significant, the data in an E_{peak}–E_{iso} plane, obtained by distinct satellites, display different distributions. Eventually, it was pointed out that the selection effects for a detector with a high threshold allow detecting only GRBs in the area where GRBs follow the Amati relation (the so called Amati region), hence these GRBs are not useful cosmological probes.

Continuing to investigate the reliability of the Amati relation, Kocevski and Petrosian (2013) reproduced this relation via a population synthesis code, which is used to mimic the prompt emission of GRBs. Under given assumptions on the distributions of spectral parameters and the luminosity function of the population, as well as the comoving rate density, the author investigated how bursts of disparate spectral features, as well as redshift, would seem to a gamma-ray detector on Earth in the observer frame. It was discovered that the Amati correlation can be reproduced, despite the initial simulations assuming only a weak association among the E_{peak} and E_{iso} variables. Thus, the left boundary of the correlation for the low-luminosity GRBs would have been produced by the limited flux detection threshold. In contrast, due to the Malmquist bias effects only the brightest GRBs are observed at large redshifts, leading to the problem of having undetected bursts of high E_{iso} and low E_{peak}, since the redshifted E_{peak} would go to energies where the detectors lose their sensitivity. For this reason, this correlation could be due to several concomitant factors, the detection threshold of the instrument, the intrinsic limit in the GRB luminosity function and the extensive redshift. Although in this treatment the model used is simplified, it is crucial to understand that selection effects can bias and undermine the reliability of observed correlations.

Instead, the main conclusion drawn from the research of Heussaff *et al* (2013) is that the E_{peak}–E_{iso} relation is generated by a physical constraint that does not allow the existence of high values of E_{iso} and low values of E_{peak}, and that the sensitivity of γ-ray and optical detectors favours GRBs located in the E_{peak}–E_{iso} plane near these constraints. These two effects seem to explain the different results obtained by several authors investigating the E_{peak}–E_{iso} relation.

Amati and Della Valle (2013), to further discuss the issue of the dependence of the Amati relation on the redshift, analyzed the reliability of the E_{peak}–E_{iso} relation using a sample of 156 GRBs available up to the end of 2012. They divided this sample into subsets with different redshift ranges (e.g. $0.1 < z < 1$, $1 < z < 2$, etc), pointing out that the selection effects are not significant because the slope, normalization and scatter of the correlation remain constant. They found

$$\log \frac{E_{\text{peak}}}{1 \text{ keV}} = 0.5 \log \frac{E_{\text{iso}}}{10^{52} \text{ erg}} + 2. \qquad (4.5)$$

Finally, Mochkovitch and Nava (2015), with a model that took into account the small number of GRBs with large E_{iso} and small E_{peak}, pointed out that the scatter of the intrinsic E_{peak}–E_{iso} relation is larger than the scatter of the observed one.

4.4 Selection effects for the isotropic luminosity

Ghirlanda *et al* (2012) studied a dataset of 46 GRBs and claimed that the flux limit—introduced to take into account selection biases related to L_{iso}—generates a constraint in the L_{iso}–E_{peak} plane. Given that this constraint corresponds to the observed relation, they pointed out that 87% of the simulations gave a statistically meaningful correlation, but only 12% returned a slope, normalization and scatter compatible with those of the original dataset. There is a non-negligible chance that a boundary with asymmetric scatter may exist due to some intrinsic features of GRBs, but to validate this hypothesis additional complex simulations would be required.

Additionally, they performed Monte Carlo simulations of the GRB population under different assumptions for their luminosity functions. Assuming there is no correlation between E_{peak} and L_{iso}, they were unable to reproduce it, thus confirming the existence of an intrinsic correlation between E_{peak} and L_{iso} at more than 2.7σ. For this reason, there should be a relation between these two parameters that does not originate from detector limits (see figure 4.2).

4.5 Selection effects for the peak luminosity

Yonetoku *et al* (2010) investigated how the truncation effects and the redshift evolution affect the L_{peak}–E_{peak} relation. They claimed that the selection bias due to truncations might occur when the detected signal is comparable to the detector threshold, and showed that the relation is indeed redshift-dependent.

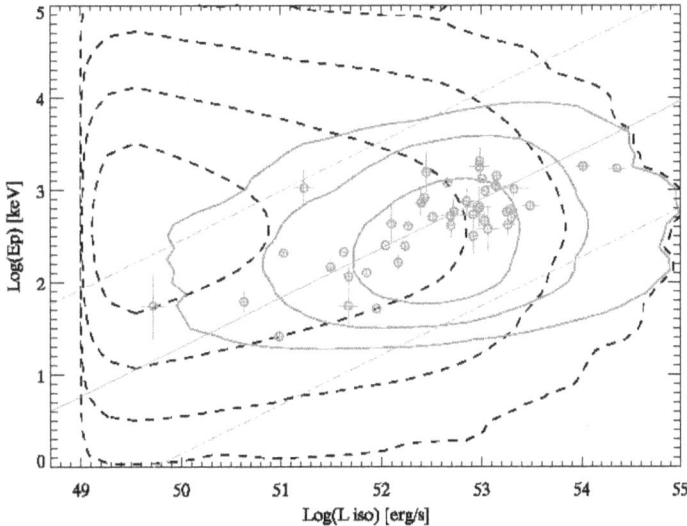

Figure 4.2. Simulation showing the variation of the density from (Ghirlanda *et al* 2012). The simulated sample is represented by the dashed contour (1-, 2-, 3- and 4σ levels). The sample of simulated GRBs with a flux greater than the constraint $F_{lim} = 2.6\,\mathrm{ph\,cm^{-2}\,s^{-1}}$ in the 15–150 keV energy band is depicted by red solid contours (1-, 2- and 3σ levels). The blue circles denote the 46 *Swift* GRBs used by Nava *et al* (2012): the solid line represents the best-fit of the L_{iso}–E_{peak} relation, and the dot-dashed lines display the 3σ region around the best-fit line. Reproduced from Ghirlanda *et al* 2012. By permission of Oxford University Press on behalf of the Royal Astronomical Society.

Shamoradi (2013) studied the E_{peak}–E_{iso} and L_{peak}–E_{peak} relations, and constructed a model describing both the luminosity function and the distribution of the prompt spectral and temporal parameters, taking into account the detection threshold of γ-ray instruments, in particular BATSE and *Fermi*/GBM. Analyzing the prompt emission data of 2130 BATSE GRBs, he demonstrated that SGRBs and LGRBs are similar in a four-dimensional space of L_{peak}, E_{iso}, E_{peak} and T_{90}. Moreover, he showed that these two relations are strongly biased by selection effects, questioning their usefulness as cosmological probes. Similar E_{peak}–E_{iso} and L_{peak}–E_{peak} relations, with analogous correlation coefficient and significance, should hold for SGRBs. Based on the multivariate log-normal distribution used to model the luminosity function, it was predicted that the strong correlation between T_{90} and both E_{iso} and L_{peak} was valid for SGRBs as well as for LGRBs.

Shahmoradi and Nemiroff (2015) investigated the luminosity function, energetics, relations among GRB prompt parameters and methodology for classifying SGRBs and LGRBs using 1931 BATSE events. Employing again the multivariate log-normal distribution model, they found out statistically meaningful L_{peak}–E_{peak} and E_{peak}–E_{iso} relations with $\rho = 0.51 \pm 0.10$ and $\rho = 0.60 \pm 0.06$, respectively.

Yonetoku *et al* (2004) and Petrosian *et al* (2015) showed how L_{peak} undergoes redshift evolution. These authors found a strong redshift evolution, $L_{peak} \propto (1 + z)^{2.0-2.3}$, an evolution which is roughly compatible in 1.5σ between the authors. Different data samples were used, but the statistical method, i.e. the Efron and Petrosian (1992) one, was the same. This tool uses a non-parametric approach, a modification of the Kendall τ statistics. A simple $f(z) = (1 + z)^{\alpha}$ or a more complex redshift function was chosen and in both cases compatible results were found for the evolution. After a proper correction of the L_{peak} evolution, Dainotti *et al* (2015a) established the intrinsic L_{peak}–L_X relation. This statistical method is very general and it can also be applied to properties of the afterglow emission. For example, Dainotti *et al* (2013) found the evolution functions for the luminosity at the end of the plateau emission, L_X, and its rest-frame duration, T_a^*, aka the Dainotti relation, and they demonstrated the intrinsic nature of the L_X–T_a^* relation. From the existence of the intrinsic nature of the L_X–T_a^* and L_{peak}–L_X relations, Dainotti *et al* (2016) discovered the extended L_X–T_a^*–L_{peak} relation, which is intrinsic as being a combination of two intrinsic correlations. However, in this paper the authors had to deal with additional selection bias problems. They analyzed two samples, the full set of 122 GRBs and the gold sample composed of 40 GRBs with high-quality data in the plateau emission. The selection criteria were defined carefully enough not to introduce any biases, which were indeed shown to be valid by a statistical comparison of the whole and gold samples. They found a tight L_a–T_a^*–L_{peak} relation, and showed that for the gold sample it has a 54% smaller scatter than a corresponding L_a–T_a^* relation obtained for the whole sample. Moreover, it was shown via the bootstrapping method that the reduction of scatter is not due to the smaller sample size, hence the 3D relation for the gold sample is intrinsic in nature, and is not biased by the selection effects related to the definition of this sample. It follows that a more careful consideration of the selection effects can provide insight into the physics underlying the GRB emission mechanism. In fact, this fundamental plane relation can be a very reliable test for physical models. An open question would be if the magnetar model (Zhang and Mészáros 2001, Troja *et al* 2007, Rowlinson *et al* 2014, Rea *et al* 2015, Stratta *et al* 2018) can still be a plausible explanation of this relation as it was for the L_X–T_a^* relation. Thus, once correlations are corrected for selection bias, they can be very good candidates to test and possibly efficiently discriminate among plausible theoretical models.

4.6 Selection effects for the lag time and the rise time

Considering τ_{lag} and its dependence on the redshift, Azzam (2012) studied the evolutionary effects of the L_{peak}–τ_{lag} relation using 19 GRBs detected by *Swift*, and found the results to be in perfect agreement with those obtained through other methods (such as the ones from Tsutsui *et al* (2008), who used redshifts obtained through the Yonetoku relation to study the dependence of the L_{peak}–τ_{lag} one on redshift). Specifically, he divided the data sample into three redshift ranges (i.e. 0.540–1.091, 1.101–1.727, 1.949–3.913) and calculated the slope and normalization of the log–log relation in each of them. In the first bin a slope of -0.92 ± 0.19 and a normalization of 51.94 ± 0.11 were found with $r = -0.89$. In the second bin the slope was -0.82 ± 0.12 and the normalization 52.12 ± 0.08 with $r = -0.94$. In the third bin the relation had a slope equal to -0.04 ± 0.22 and the normalization was 52.90 ± 0.12 with $r = -0.06$. Therefore, the L_{peak}–τ_{lag} relation seems to evolve with redshift, however, this conclusion is only tentative since there is the problem that each redshift range is not equally populated, and is limited by low statistics and by the significant scatter in the relation. Therefore, the L_{peak}–τ_{lag} relation is redshift-dependent, but this result is not conclusive due to the paucity of the sample and significant scatter.

Kocevski and Petrosian (2013) found that in individual pulses the observer-frame cosmological time dilation is masked out because only the most luminous part of the light curve can be observed by GRB detectors. Therefore, the duration and E_{iso} for GRBs close to the detector threshold need to be considered as lower limits, and the temporal characteristics are not sufficient to discriminate between LGRBs and SGRBs (see also Tarnopolski (2015) for a novel and successful attempt to use for this purpose non-standard parameters via machine learning).

Considering instead the rise time, Wang *et al* (2011) used 72 LGRBs observed by *BeppoSAX* and *Swift* to study the L_{iso}–τ_{RT} relation, and found that the relation is not dependent on the redshift. In fact, for the total sample they obtained

$$\log \frac{L_{\text{iso}}}{1 \text{erg s}^{-1}} = (52.68 \pm 0.07) - (1.12 \pm 0.14) \log \frac{\tau_{\text{RT}}^*}{0.1 \text{ s}}, \qquad (4.6)$$

with $\sigma_{\text{int}} = 0.48 \pm 0.05$. Additionally, dividing the dataset into four redshift bins (i.e. 0–1, 1–2, 2–3 and 3–8.5), the slope and normalization of this relation remained nearly constant. This represents good evidence for the L_{iso}–τ_{RT} relation not being influenced by evolutionary effects.

References

Amati L 2006 The $E_{p,i}$–E_{iso} correlation in gamma-ray bursts: updated observational status, re-analysis and main implications *Mon. Not. R. Astron. Soc.* **372** 233–45

Amati L and Della Valle M 2013 Measuring cosmological parameters with gamma ray bursts *Int. J. Mod. Phys.* D **22** 1330028

Amati L *et al* 2002 Intrinsic spectra and energetics of *BeppoSAX* gamma-ray bursts with known redshifts *Astron. Astrophys.* **390** 81–9

Amati L, Guidorzi C, Frontera F, Della Valle M, Finelli F, Landi R and Montanari E 2008 Measuring the cosmological parameters with the $E_{p,i}$–E_{iso} correlation of gamma-ray bursts *Mon. Not. R. Astron. Soc.* **391** 577–84

Azzam W J 2012 Dependence of the GRB lag–luminosity relation on redshift in the source frame *Int. J. Astron. Astrophys.* **2** 1–5

Band D L and Preece R D 2005 Testing the gamma-ray burst energy relationships *Astrophys. J.* **627** 319–23

Bosnjak Z, Celotti A, Longo F and Barbiellini G 2005 Energetics—spectral correlations vs the BATSE gamma-ray bursts population, arXiv:astro-ph/0502185

Bošnjak Ž, Götz D, Bouchet L, Schanne S and Cordier B 2014 The spectral catalogue of *INTEGRAL* gamma-ray bursts. results of the joint IBIS/SPI spectral analysis *Astron. Astrophys.* **561** A25

Butler N R, Kocevski D and Bloom J S 2009 Generalized tests for selection effects in gamma-ray burst high-energy correlations *Astrophys. J.* **694** 76–83

Collazzi A C, Schaefer B E and Moree J A 2011 The total errors in measuring E_{peak} for gamma-ray bursts *Astrophys. J.* **729** 89

Collazzi A C, Schaefer B E, Goldstein A and Preece R D 2012 A significant problem with using the Amati relation for cosmological purposes *Astrophys. J.* **747** 39

Dainotti M G, Petrosian V, Singal J and Ostrowski M 2013 Determination of the intrinsic luminosity time correlation in the x-ray afterglows of gamma-ray bursts *Astrophys. J.* **774** 157

Dainotti M, Petrosian V, Willingale R, O'Brien P, Ostrowski M and Nagataki S 2015a Luminosity–time and luminosity–luminosity correlations for GRB prompt and afterglow plateau emissions *Mon. Not. R. Astron. Soc.* **451** 3898–908

Dainotti M G, Del Vecchio R, Nagataki S and Capozziello S 2015b Selection effects in gamma-ray burst correlations: consequences on the ratio between gamma-ray burst and star formation rates *Astrophys. J.* **800** 31

Dainotti M G, Postnikov S, Hernandez X and Ostrowski M 2016 A fundamental plane for long gamma-ray bursts with x-ray plateaus *Astrophys. J.* **825** L20

Dichiara *et al* 2013 A search for pulsations in short gamma-ray bursts to constrain their progenitors *Astrophys. J.* **777** 132

Efron B and Petrosian V 1992 A simple test of independence for truncated data with applications to redshift surveys *Astrophys. J.* **399** 345–52

Efron B and Petrosian V 1999 Nonparametric methods for doubly truncated data *J. Am. Stat. Assoc.* **94** 824–34

Ford L A *et al* 1995 BATSE observations of gamma-ray burst spectra. 2: peak energy evolution in bright, long bursts *Astrophys. J.* **439** 307–21

Frontera F, Amati L, Guidorzi C, Landi R and in't Zand J 2012 Broadband time-resolved $E_{p,i}$–L_{iso} correlation in gamma-ray bursts *Astrophys. J.* **754** 138

Ghirlanda G, Ghisellini G and Celotti A 2004a The spectra of short gamma-ray bursts *Astron. Astrophys.* **422** L55–8

Ghirlanda G, Ghisellini G and Lazzati D 2004b The collimation-corrected gamma-ray burst energies correlate with the peak energy of their νF_{ν} spectrum *Astrophys. J.* **616** 331–8

Ghirlanda G, Ghisellini G, Firmani C, Celotti A and Bosnjak Z 2005a The peak luminosity–peak energy correlation in gamma-ray bursts *Mon. Not. R. Astron. Soc.* **360** L45–9

Ghirlanda G, Ghisellini G, Lazzati D and Firmani C 2005b The updated E_{peak}–E_{g} correlation in GRBs *Nuovo Cim.* C **28** 303

Ghirlanda G, Nava L, Ghisellini G, Firmani C and Cabrera J I 2008 The E_{peak}–E_{iso} plane of long gamma-ray bursts and selection effects *Mon. Not. R. Astron. Soc.* **387** 319–30

Ghirlanda G, Nava L and Ghisellini G 2010 Spectral–luminosity relation within individual *Fermi* gamma rays bursts *Astron. Astrophys.* **511** A43

Ghirlanda G *et al* 2012 The impact of selection biases on the E_{peak}–L_{iso} correlation of gamma-ray bursts *Mon. Not. R. Astron. Soc.* **422** 2553–9

Goldstein A, Preece R D and Briggs M S 2010 A new discriminator for gamma-ray burst classification: the E_{peak}–fluence energy ratio *Astrophys. J.* **721** 1329–32

Heussaff V, Atteia J-L and Zolnierowski Y 2013 The E_{peak}–E_{iso} relation revisited with *Fermi* GRBs. Resolving a long-standing debate? *Astron. Astrophys.* **557** A100

Kaneko Y, Preece R D, Briggs M S, Paciesas W S, Meegan C A and Band D L 2006 The complete spectral catalog of bright BATSE gamma-ray bursts *Astrophys. J. Suppl. Ser.* **166** 298–340

Kocevski D and Petrosian V 2013 On the lack of time dilation signatures in gamma-ray burst light curves *Astrophys. J.* **765** 116

Lee T T and Petrosian V 1996 Distributions of peak flux and duration for gamma-ray bursts *Astrophys. J.* **470** 479

Lloyd N M and Petrosian V 1999 Distribution of spectral characteristics and the cosmological evolution of gamma-ray bursts *Astrophys. J.* **511** 550–61

Lloyd N M, Petrosian V and Mallozzi R S 2000 Cosmological versus intrinsic: the correlation between intensity and the peak of the $\nu F_{\frac{1}{2}}$ spectrum of gamma-ray bursts *Astrophys. J.* **534** 227–38

Lu R-J, Wei J-J, Liang E-W, Zhang B-B, Lü H-J, Lü L-Z, Lei W-H and Zhang B 2012 A comprehensive analysis of *Fermi* gamma-ray burst data. II. E_p evolution patterns and implications for the observed spectrum–luminosity relations *Astrophys. J.* **756** 112

Mallozzi R S, Paciesas W S, Pendleton G N, Briggs M S, Preece R D, Meegan C A and Fishman G J 1995 The νF_{ν} peak energy distributions of gamma-ray bursts observed by BATSE *Astrophys. J.* **454** 597

Mochkovitch R and Nava L 2015 The E_p–E_{iso} relation and the internal shock model *Astron. Astrophys.* **577** A31

Nakar E and Piran T 2005 Outliers to the peak energy–isotropic energy relation in gamma-ray bursts *Mon. Not. R. Astron. Soc.* **360** L73–6

Nava L *et al* 2012 A complete sample of bright *Swift* long gamma-ray bursts: testing the spectral–energy correlations *Mon. Not. R. Astron. Soc.* **421** 1256–64

Petrosian V, Kitanidis E and Kocevski D 2015 Cosmological evolution of long gamma-ray bursts and the star formation rate *Astrophys. J.* **806** 44

Petrosian V, Singal J and Stawarz L 2013 Luminosity correlations, luminosity evolutions, and radio loudness of AGNs from multiwavelength observations *Proc. Int. Astron. Union* **9** 172

Rea N, Gullón M, Pons J A, Perna R, Dainotti M G, Miralles J A and Torres D F 2015 Constraining the GRB–magnetar model by means of the galactic pulsar population *Astrophys. J.* **813** 92

Rowlinson A, Gompertz B P, Dainotti M, O'Brien P T, Wijers R A M J and van der Horst A J 2014 Constraining properties of GRB magnetar central engines using the observed plateau luminosity and duration correlation *Mon. Not. R. Astron. Soc.* **443** 1779–87

Shahmoradi A and Nemiroff R J 2015 Short versus long gamma-ray bursts: a comprehensive study of energetics and prompt gamma-ray correlations *Mon. Not. R. Astron. Soc.* **451** 126–43

Shamoradi A 2013 A multivariate fit luminosity function and world model for long gamma-ray bursts *Astrophys. J.* **766** 111

Stratta G, Dainotti M G, Dall'Osso S, Hernandez X and De Cesare G 2018 On the magnetar origin of the GRBs presenting X-ray afterglow plateaus *Astrophys. J.* **869** 155

Tarnopolski M 2015 Analysis of *Fermi* gamma-ray burst duration distribution *Astron. Astrophys.* **581** A29

Troja E *et al* 2007 *Swift* observations of GRB 070110: an extraordinary x-ray afterglow powered by the central engine *Astrophys. J.* **665** 599–607

Tsutsui R, Nakamura T, Yonetoku D, Murakami T, Tanabe S and Kodama Y 2008 Redshift dependent lag–luminosity relation in 565 BASTE gamma ray bursts *AIP Conf. Ser.* **1000** 28–31

Wang F-Y, Qi S and Dai Z-G 2011 The updated luminosity correlations of gamma-ray bursts and cosmological implications *Mon. Not. R. Astron. Soc.* **415** 3423–33

Yonetoku D, Murakami T, Nakamura T, Yamazaki R, Inoue A K and Ioka K 2004 Gamma-ray burst formation rate inferred from the spectral peak energy–peak luminosity relation *Astrophys. J.* **609** 935–51

Yonetoku D, Murakami T, Tsutsui R, Nakamura T, Morihara Y and Takahashi K 2010 Possible origins of dispersion of the peak energy–brightness correlations of gamma-ray bursts *Publ. Astron. Soc. Jpn.* **62** 1495

Zhang B and Mészáros P 2001 Gamma-ray burst afterglow with continuous energy injection: signature of a highly magnetized millisecond pulsar *Astrophys. J.* **552** L35–8

Zhang B and Mészáros P 2002 An analysis of gamma-ray burst spectral break models *Astrophys. J.* **581** 1236–47

Chapter 5

Redshift estimators and cosmology for prompt relations

5.1 Redshift estimator for correlations between prompt parameters

Given that the value of the redshift z for the majority of GRBs is not measured, recovering a correlation able to deduce the GRB distance from parameters independent of z would improve our knowledge of GRB features. In addition, it would provide some clues on GRB distance. Here, some emblematic cases regarding the correlations used to build a redshift indicator between prompt parameters (Atteia 2003, Yonetoku *et al* 2004, Tsutsui *et al* 2013) are described. To find pseudo-redshifts of 17 *BeppoSAX* GRBs, Atteia (2003) investigated the E_{peak}–E_{iso} correlation, given that these parameters are dependent on the luminosity distance. Defining the quantity $X = \frac{n_\gamma \sqrt{T_{90}}}{E_{peak}}$, with n_γ the amount of observed photons, the theoretical evolution of this quantity was given by:

$$X = A \times f(z), \tag{5.1}$$

with A a constant and $f(z)$ is the redshift evolution in the case of a 'standard' GRB in a ΛCDM model ($H_0 = 65 \, \mathrm{km s^{-1} \, Mpc^{-1}}$, $\Omega_M = 0.3$, $\Omega_\Lambda = 0.7$). Namely, the 'standard' GRBs are defined as GRBs with parameters of the Band function $\alpha = -1.0$, $\beta = -2.3$ and $E_0 = 250$ keV. Inverting the relation of theoretical evolution, a redshift indicator was obtained through the observables. This redshift estimator is represented by the formula $z = \frac{1}{A} f^{-1}(X)$. This indicator may be able to make a quick classification of GRBs at $z > 3$, a comparison at a statistical level of the distance distributions of several GRB groups, and the evaluation of the high-z SFR.

With 689 highly luminous BATSE LGRBs, Yonetoku *et al* (2004) studied their spectra through the Band function. To retrieve significant S/N, the lowest value for the flux was established as $F_{lim} = 2 \times 10^{-7} \mathrm{erg \, cm^{-2} s^{-1}}$. Furthermore, they computed

F_{peak} and E_{peak}^{*}. With * we denote the rest-frame quantities. Then, considering the dependence of these quantities on $D_L(z, \Omega_M, h)$, they inverted the equation,

$$\log L_{\text{peak}} = (47.37 \pm 0.37) + (2.0 \pm 0.2) \log E_{\text{peak}}^{*}. \qquad (5.2)$$

and the GRBs' pseudo-redshifts were recovered.

In addition, Tsutsui *et al* (2013) studied the E_{peak}–E_{iso} and the L_{peak}–E_{peak} correlations for 71 BATSE SGRBs. From their analysis, the L_{peak}–E_{peak} correlation appears stronger than the E_{peak}–E_{iso} correlation, thus the first correlation seems to be a better redshift indicator. Considering a different form of the L_{peak}–E_{peak} correlation,

$$\frac{D_L^2(z, \Omega_M, \Omega_\Lambda)}{(1 + z)^{1.59}} = \frac{10^{52.29}}{4\pi F_{\text{peak}}} \left(\frac{E_{\text{peak}}}{774.5}\right)^{1.59}, \qquad (5.3)$$

and $(\Omega_M, \Omega_\Lambda) = (0.3, 0.7)$, they computed the pseudo-redshifts from $D_L(z, \Omega_M, \Omega_\Lambda)$ (see figure 5.1). The application of the LT correlation as a redshift estimator was studied by Dainotti *et al* (2011). They concluded that reliable outcomes can be achieved only in the case where the scatter of the correlation is less than 20% and the errors in the parameters of the analysed GRBs are small. Given its smaller scatter, better outcomes for this redshift estimator are anticipated from the recent 3D correlation (Dainotti et al 2016).

To investigate its use as redshift estimator, the hardness–intensity correlation, namely the Golenetskii correlation, between the GRB spectral peak energy and its instantaneous luminosity, was analysed by Burgess (2015), applying a hierarchical Bayesian analysis. Employing a power-law form for the correlation, the indices were

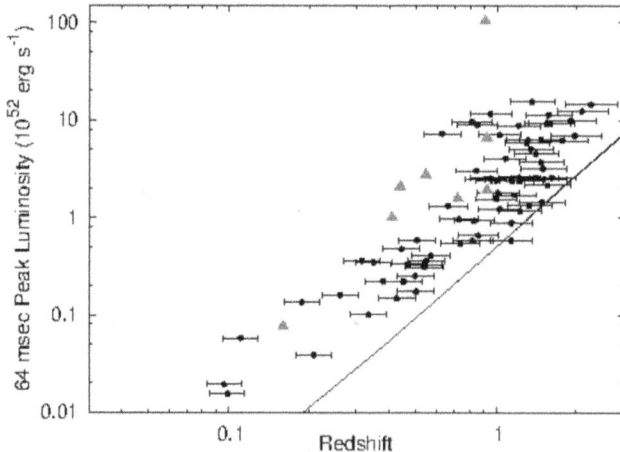

Figure 5.1. The distribution of z through the $\log L_{\text{peak}}$–$\log E_{\text{peak}}$ correlation for 71 bright BATSE SGRBs from Ghirlanda *et al* (2009). The pseudo-redshifts and L_{peak} are indicated by black dots, and secure SGRBs are displayed by red filled triangles; $z \in (0.097, 2.581)$ with $\langle z \rangle = 1.05$, while for *Swift* LGRBs $\langle z \rangle \approx 2.16$. The flux limit of $F_{\text{peak}} = 10^{-6} \text{erg cm}^{-2}\text{s}^{-1}$ is shown by the solid line. (Reproduced from Tsutsui *et al* (2013). By permission of Oxford University Press on behalf of the Royal Astronomical Society.)

found around a specific value, with scatter larger than those previously obtained, however. In addition, the rest-frame normalization of the correlation spans between 10^{51} and 10^{53} erg s^{-1}, suggesting the presence of several quantities related to the radiation, such as the strength of the magnetic field, the number of emitting electrons, the photospheric radius, the viewing angle, and so forth. Therefore, it was concluded that the hardness–intensity correlation cannot be a reliable redshift estimator. Nevertheless, an improved estimation of the correlation properties in the rest-frame is retrieved through the Bayesian analysis applied here, providing tighter constraints for the physics of the radiation models.

Guiriec *et al* (2013, 2016, 2016) investigated the use of a correlation similar to the Yonetoku one (Yonetoku *et al* 2004) as a redshift indicator. This correlation, between the non-thermal component luminosity, L_i^{nTh}, and its rest-frame spectral peak energy, $E_{\mathrm{peak,i}}^{\mathrm{nTh,rest}}$, allowed for the evaluation of three redshifts.

An additional issue of this correlation is the measurement of the intrinsic scatter. Andreon (2013) investigated this problem as well as the controversial features of the data in astronomy, such as upper/lower limits, selection biases, intrinsic scatter, the effect of the Malmquist bias on the samples, the non-Gaussianity aspect of the observational data sample, and so forth. It was claimed that the Bayesian analysis is a good choice for the description of the data characteristics. Indeed, after the data analysis, the posterior probability distribution provides all the available pieces of information about the analysed variables, and then the Monte Carlo simulations are employed for the numerical evaluation. As a main result, Andreon (2013) provided a way to bound the cosmological parameters computed through a sample of SNe and to confirm the functional form of the fit.

5.2 Cosmology

The distribution of the distance modulus $\mu(z)$, which is the difference between the apparent (m) and the absolute (M) magnitudes, versus z of SNe Ia is called the Hubble diagram (HD). This can be really helpful for the study of DE. It was already established in the literature that $\mu(z)$ is related to $D_L(z, \Omega_M, h)$:

$$\mu(z) = 25 + 5 \times \log D_L(z, \Omega_M, h), \qquad (5.4)$$

with $D_L(z, \Omega_M, h)$ associated with distinct DE EoS.

5.2.1 The problem of calibration

The circularity problem is one of the main difficulties regarding the investigation of GRB correlations for cosmological purposes. The circularity problem implies that to calculate $D_L(z, \Omega_M, h)$ a cosmological model has to be assumed *a priori*. The main reason for this issue is that GRBs are not observed at $z < 0.01$, where there would not be any dependence on the cosmological model. Indeed, the only exception to this is GRB 980425 observed at $z = 0.0085$. To solve this problem, some solutions were proposed: (a) to consider a significant sample of low-z GRBs to calibrate these correlations (indeed at $z \leqslant 0.1$ Ω_M and Ω_Λ, for a certain H_0 between 65 and 72, do not affect $D_L(z, \Omega_M, h)$); (b) to describe the observed 2D correlations taking into account a well-established theoretical model to determine their slopes and normalizations without any dependence on the cosmological model—unfortunately, this plan does not seem feasible; and (c) to calibrate the standard candles employing GRBs in a small redshift interval (Δz) around a particular value of the redshift, z_c. Now, a few ways of avoiding this circularity problem by considering some correlations between prompt parameters are presented. Even in the case of correlations between afterglow parameters or prompt–afterglow parameters, the circularity problem can be managed in the same way.

Liang and Zhang (2006) outlined a new GRB luminosity indicator, dissimilar from earlier ones, given by $E_{\mathrm{iso}} = aE_{\mathrm{peak}}^{b_1}T_{O,a}^{b_2}$. From the analysis of this luminosity indicator it was found that a is dependent on the cosmological model assumed. This is not true for b_1 and b_2 until Δz is adequately small (see figure 5.2). Based on how large the GRB sample is and the errors on the observables, Δz can be selected appropriately. As claimed by Wang et al (2011) and Wang et al (2015), the most convenient method is to group GRBs within narrow redshift intervals centred at a particular z_c ($z_c \sim 1$ or $z_c \sim 2$, the range where the GRB z distribution has its maximum).

Employing the log E_{peak}–log E_γ correlation (Ghirlanda et al 2004), Ghirlanda et al (2006) also built a luminosity indicator given by $E_{\mathrm{peak}} = a \times E_\gamma^b$.

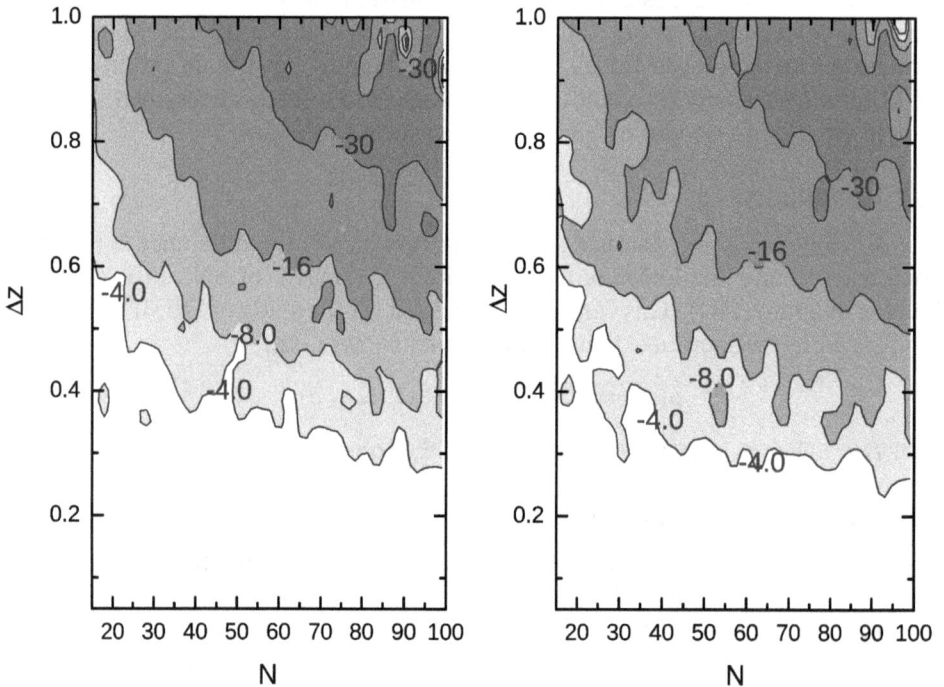

Figure 5.2. The (N, Δz) space with the log P distribution. The regions where b_1 and b_2 are statistically dependent on Ω_M (P $<10^{-4}$) are indicated by grey contours. The appropriate area for the calibration procedure is represented by the white zone. (Reproduced from Liang and Zhang 2006b. By permission of Oxford University Press on behalf of the Royal Astronomical Society.)

Using a dataset of 19 GRBs observed by *BeppoSAX* and *Swift*, they estimated the smallest number of GRBs (N) needed to calibrate the correlation. Considering N GRBs in the range Δz (distributed around z_c) they fitted the correlation for several values of Ω_M and Ω_Λ. If the slope, b, changed by less than 1% the correlation is calibrated. With N, Δz and z_c free parameters, distinct z_c and z intervals, $\Delta z \in (0.05, 0.5)$, were tested through Monte Carlo simulations. It was found that at each z the smaller the N the greater the slope variation, Δb, for equal values of Δz. The main reason for this finding is that in this case the scatter of the correlation is large. On the other hand, for higher z_c a narrower Δz is necessary to obtain tiny Δb. They pointed out that 12 GRBs with z in the range (0.9, 1.1) are enough for the calibration of the log E_{peak}–log E_γ correlation. At $z_c = 2$ a smaller redshift bin is required, such as z in in the range (1.95, 2.05). However, given that the GRB sample is not wide enough, this procedure may be not favourable.

Alternatively, it is possible to assume that $\mu(z)$ of a GRB at redshift z has the same value as that of an SN Ia at the same redshift. Thus, SNe Ia can be used as distance estimators for a calibration independent of the cosmological model assumed. In this procedure GRBs are filling the gap beyond SNe Ia, reaching very high redshift and providing an extended distance indicator. Thus, taking the value of $\mu(z)$ from the SN Ia HD, the value for GRBs at $z \leqslant 1.4$ can also be inferred. Then, this value can be used to calibrate the correlations at higher z (Kodama *et al* 2008, Liang *et al* 2008, Wei and Zhang 2009). $\mu(z)$ for SNe Ia is expressed by the equation

$$
\begin{aligned}
\mu(z) &= 25 + (5/2)(\log y - k) \\
&= 25 + (5/2)(a + b \log x - k),
\end{aligned}
\tag{5.5}
$$

with $y = k D_L^2(z, \Omega_M, h)$ where k is a constant independent of the redshift, and a and b are the correlation parameters.

Considering this calibration independent of the redshift, the HD at higher z can be computed through the calibrated correlations.

The light curves of eight LGRBs with associated SNe were studied by Li and Hjorth (2014), pointing out a correlation between the peak magnitude and the decay rate at 5, 10 and 15 days as observed in SNe Ia. However, comparing this with the well-known (Phillips 1993) correlation for SNe Ia, it was concluded that the SNe associated with GRBs and SNe Ia have two distinct progenitors. In addition, this outcome gave the possibility of employing GRBs associated with SNe as plausible standard candles. Furthermore, the optical light curves of eight LGRBs associated with SNe (six light curves are in common with the ones used by Li and Hjorth (2014)) were analysed by Cano (2014). They pointed out a correlation between their luminosity and the width of the GRB light curves in relation to the SN 1998bw template. This finding also indicated that GRBs associated with SNe could be employed as standard candles.

5.2.2 Applications of GRB correlations between prompt parameters to cosmology

To constrain the parameters of the cosmological model through a given sample of GRBs, Dai *et al* (2004) and Xu *et al* (2005) investigated a procedure based on the

E_{peak}–E_γ correlation using 12 and 17 GRBs, respectively. Dai *et al* (2004) recovered $\Omega_M = 0.35^{+0.15}_{-0.15}$, while Xu *et al* (2005) found $\Omega_M = 0.15^{+0.45}_{-0.13}$, within 1σ confidence level (CL). These results are in agreement with those for SNe Ia.

Ghirlanda *et al* (2006), studying a sample of 19 GRBs, concluded that E_{peak}–E_γ and E_{peak}–E_{iso}–T_{break} correlations can be employed for cosmological purposes for the homogeneous (HM, see the left panel in figure 5.3) and wind circumburst medium (WM, see the middle panel in figure 5.3) events. Later, Ghirlanda (2009), using an updated sample of 29 GRBs, confirmed earlier outcomes (see the right panel in figure 5.3).

Ghirlanda *et al* (2006) employed three distinct procedures for fitting the cosmological parameters, using a sample of GRBs, to avoid the circularity problem:

1. The scatter procedure in which the cosmological parameters, such as Ω_M and Ω_Λ is required to be constrained. These parameters are selected to fit the correlation. In this procedure a χ^2 surface dependent on these quantities is retrieved. The minimum value of this surface gives the best cosmological model.

2. The luminosity distance procedure is divided into few stages: (1) select the cosmological parameters to fit the E_{peak}–E_γ correlation; (2) retrieve the best value of $\log E_\gamma$; (3) compute $\log E_{iso}$ from $\log E_\gamma$ and consequently $D_L(z, \Omega_M, \Omega_\Lambda)$; (4) compare $D_L(z, \Omega_M, \Omega_\Lambda)$ with that from the cosmological model to compute χ^2. After several selections of the values of the cosmological parameters, a χ^2 surface is recovered. As in the previous procedure, the best cosmological model is provided by the minimum χ^2.

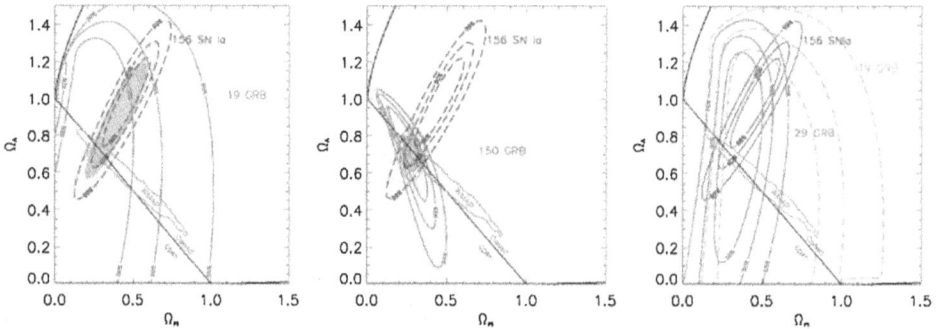

Figure 5.3. Left panel: The 68.3%, 90% and 99% contours of Ω_M and Ω_Λ from the E_{peak}–E_γ correlation for a dataset of 19 GRBs and HM (Ghirlanda *et al* 2006) are indicated by the solid red line. The red cross represents the centre of these areas. The smallest χ^2 is provided by this point, together with values $\Omega_M = 0.23$, $\Omega_\Lambda = 0.81$. (Reproduced from Ghirlanda *et al* (2006).) Middle panel: as in the left panel but retrieved through 150 simulated GRBs and the E_{peak}–E_γ correlation for the WM. (Reproduced from Ghirlanda *et al* (2006). © 2006 IOP Publishing Ltd and Deutsche Physikalische Gesellschaft.) In both panels, the dashed blue lines indicate the contours for 156 SNe Ia from the 'Gold' dataset by Riess *et al* (2004). The shaded contours display the boundaries retrieved from GRBs and SNe together. In this picture, 90% contours recovered through a WMAP sample are also plotted. Right panel: as in the previous panels, but with data extended to January 2009. The solid line indicates the updated sample of 29 GRBs, while the dashed line represents the earlier dataset of 19 GRBs. The blue thin line displays the boundaries for a sample of 156 SNe Ia by Riess *et al* (2004), while the green line depicts the constraints for a WMAP dataset. (Reproduced with permission from Ghirlanda (2009). Copyright 2009 AIP Publishing.)

3. The Bayesian procedure: the first two procedures rely on the statement that the origin of some correlations, e.g. the E_{peak}–E_γ, relies on the physical quantities characterizing them. However, these procedure do not explain the fact that the GRB physics also influences the correlations and, for this reason, they may be unique. A more composite procedure ground on the Bayesian analysis was developed by Firmani *et al* (2005) using a sample of both GRBs and SNe Ia.

The E_{peak}–E_{iso} correlation can be used for computing Ω_M, as shown in Amati *et al* (2008). The scatter of this correlation can be properly estimated through the maximum likelihood method, providing in the case of a flat universe $\Omega_M = 0.04$–0.40 (68% CL) with a best value of $\Omega_M = 0.15$. The value $\Omega_M = 1$ is ruled out at >99.9% CL. In this case the circularity problem (see section 5.2.1) does not influence the results because no particular statements on the E_{peak}–E_{iso} correlation are asserted. For computing the normalization, further calibrators are not employed. In addition, their findings are independent of those from a sample of SNe Ia. Due to the expectations of the present and forthcoming space missions the errors in Ω_M and Ω_Λ can be considerably diminished.

To analyse the constraints on Ω_M and Ω_Λ, Tsutsui *et al* (2009a) applied three correlations to 31 low-z GRBs and 29 high-z GRBs. The first correlation was E_{peak}–T_L–L_{peak}, the second was the L_{peak}–E_{peak} and the third was E_{peak}–E_{iso}. The last two correlations were calibrated through GRBs at $z \leqslant 1.8$ (see the upper left and upper middle panels in figure 5.4, respectively). In the case of a ΛCDM model with $\Omega_k = \Omega_M + \Omega_\Lambda - 1$, and Ω_k the density of the spatial curvature, the constraints for the L_{peak}–E_{peak} and E_{peak}–E_{iso} correlations are incompatible within 1σ, but comparable in 2σ. For this reason, they defined the quantity $T_L = E_{iso}/L_{peak}$, and the following correlation was found:

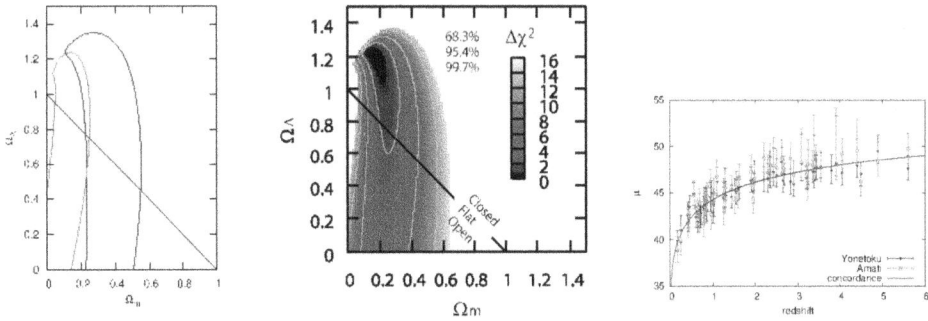

Figure 5.4. Upper left panel: 68.3% CL contours for Ω_M and Ω_Λ recovered through the Amati (presented in red) and the Yonetoku (indicated in blue) correlations. The flat universe is displayed by a black solid line. The constraints are in agreement within 2σ. (Reproduced Tsutsui *et al* (2009a). Copyright 2009 IOP Publishing.) Upper middle panel: contours for Ω_M and Ω_Λ recovered through the E_{peak}–T_L–L_{peak} correlation. (Reproduced from Tsutsui *et al* (2009a). Copyright 2009 IOP Publishing.) Upper right panel: HD obtained considering the Amati (presented in red) and the Yonetoku (indicated in blue) correlations. At high z a systematic discrepancy of HD, not observed at low z, shows up. (Reproduced from Tsutsui *et al* (2009a). Copyright 2009 IOP Publishing.)

$$\log L_{\text{peak}} = (49.87 \pm 0.19) + (1.82 \pm 0.08) \log E_{\text{peak}} - (0.34 \pm 0.09) \log T_L. \quad (5.6)$$

With the introduction of this quantity, the uncertainty of this correlation presented a decrease of 40%. For GRBs at $1.8 < z < 5.6$ the correlation found $(\Omega_M, \Omega_\Lambda) = (0.17^{+0.15}_{-0.08}, 1.21^{+0.07}_{-0.61})$, in agreement with the ΛCDM model (see the upper right panel in figure 5.4).

To stretch out the HD to $z = 5.6$, Tsutsui et al (2009b) analysed the L_{peak}–E_{peak} correlation using 63 GRBs and 192 SNe Ia (see figure 5.5). They applied three cosmological models: the ΛCDM model, a non-dynamical DE model (with the parameter of the DE EoS, $w_a = 0$) and a dynamical DE model (with the DE EoS $w(z) = w_0 + w_a z/(1 + z)$). Their results obtained with the GRB sample were compatible with the ΛCDM model ($\Omega_M = 0.28$, $\Omega_\Lambda = 0.72$, $w_0 = -1$, $w_a = 0$) at 2σ CL. Then, they employed Monte Carlo simulations to compute the boundaries of the DE EoS parameters predicted by *Fermi* and *Swift* detections. It was found that with a further 150 GRBs these boundaries should become considerably tighter.

Later, Tsutsui et al (2010) claimed that $D_L(z, \Omega_M, \Omega_\Lambda)$ can be computed with an uncertainty of $\sim$16% through the E_{peak}–T_L–L_{peak} correlation. This level of precision can be advantageous for investigating the DE at $z > 3$. Furthermore, Wang et al (2011) claimed that correlations such as LT (Dainotti et al 2008) and $L_{X,p}$–$T^*_{X,p}$ (Qi and Lu 2010) can be regarded as standard candles after suitable calibration. Wang et al (2011) showed that this result can be achieved through the minimization of χ^2, obtained by the maximum likelihood procedure, over both the correlation coefficients and the cosmological parameters. Indeed, through these correlations, the cosmic acceleration at high redshift could be better established and, simultaneously, the constraints on the cosmic acceleration at low redshift may be enhanced. To differentiate between DE and modified gravity models, GRBs are in principle good tools to determine the cosmological parameters (Wang et al 2009, Wang and Dai 2009, Vitagliano et al 2010, Capozziello and Izzo 2008).

With a dataset of 120 GRBs, Amati (2012) and Amati and Della Valle (2013) continued the work by Amati et al (2008). To examine the DE, they simulated a dataset of 250 GRBs to recover the 68% CL contours in the Ω_M–Ω_Λ space. Then, these contours were compared to those from SNe Ia, CMB and galaxy clusters.

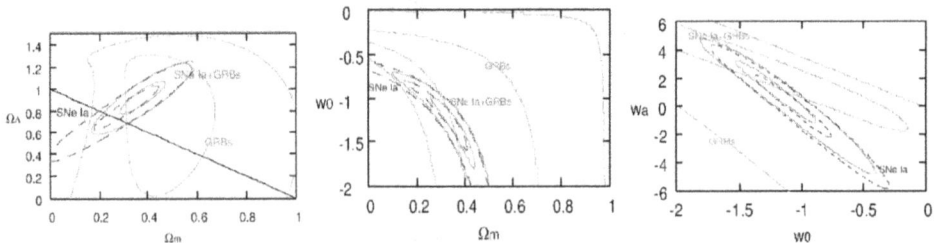

Figure 5.5. Maximum likelihood $\Delta\chi^2$ contours, at 68.3% and 99.7% CL, for the $(\Omega_M, \Omega_\Lambda)$ (left panel), the (Ω_M, w_0) (middle panel) and the (w_0, w_a) (right panel) distributions. Light blue dash-dotted lines indicate the constraints for GRBs, the blue dotted lines the constraints for SNe Ia, and the red solid lines the constraints for SNe Ia and GRBs together. The flat Universe is indicated by a black solid line. (Reproduced from Tsutsui et al (2009b). By permission of Oxford University Press on behalf of the Royal Astronomical Society.)

Amati and Della Valle (2013) applied a Monte Carlo method to simulate the sample. They considered the observed z distribution of the sample, the coefficients and the scatter of the E_{peak}–E_{iso} correlation, and the errors in E_{peak} and E_{iso} quantities. From the simulation it was found that $\approx$250 GRBs are enough to recover Ω_M with a precision in agreement with that from SNe Ia. They also predicted values of Ω_M and w_0 from present and forthcoming detections. It was claimed that the E_{peak}–E_{iso} correlation is calibrated with a precision of 10%. Indeed, they employed $D_L(z, \Omega_M, \Omega_\Lambda)$ obtained from a sample of SNe Ia and a GRB sample wide enough in a narrow interval ($\Delta z \sim 0.1$–0.2), to self-calibrate the GRB correlation. Finally, they found that the precision and reliability of the self-calibration of the E_{peak}–E_{iso} correlation may improve for a larger sample of GRBs in each redshift bin.

Lin et al (2016) separated the dataset of 116 GRBs, with z between 0.17 and 8.2 and used by Wang et al (2011) into two subsamples: GRBs with $z < 1.4$ and GRBs with $z > 1.4$. The main aim was to analyse the redshift dependence of the L_{peak}–τ_{lag}, L_{peak}–V, E_{peak}–L_{peak}, E_{peak}–E_γ, L_{iso}–τ_{RT} and E_{peak}–E_{iso} correlations. From their analysis, GRBs at $z < 1.4$ found a E_{peak}–E_γ correlation compatible in 1σ CL with the one for GRBs at $z > 1.4$. The substantial value of the scatter for the L_{peak}–V did not allow for any definitive statement. In the case of the other correlations, GRBs at $z < 1.4$ and those at $z > 1.4$ are incompatible at more than 3σ CL. For this reason, GRBs were calibrated, regardless of the model assumed, through the E_{peak}–E_γ correlation. $\Omega_M = 0.302 \pm 0.142$ was recovered for GRBs at $z > 1.4$ at 1σ CL in the ΛCDM model. This result was completely compatible with the Planck 2015 results (Planck Collaboration et al 2016). To summarize, an independent estimation of Ω_M has already been supplied by GRBs, and it was supported by simulations that a precision similar to that of SNe Ia can be accomplished by GRBs in the future.

5.3 Statistical approaches related to SN Ia cosmology

In this section, works on the cosmology of SNe Ia and the current status of the field are studied to improve our knowledge of the cosmological usage of GRBs. Indeed, some difficulties with the statistical procedures for the analysis of SNe Ia are presented in the literature, even though they are well-established standard candles. For this reason, in the case of GRBs, physical phenomena which are considerably less well understood than SNe Ia, the errors of the GRB correlations are significant, with the selection biases influencing the outcomes. In addition, the dataset of the luminosity indicators is much smaller. Statistical methods and procedures of standardization for SNe Ia and GRBs are described. In particular, the methods for SNe Ia that could be more useful and efficient for cosmological studies of GRBs are discussed.

Considering random selection biases, Mandel et al (2009) developed a description of the near-infrared (NIR) light curves of SNe Ia. This seemed a very encouraging procedure, because NIR radiation is not affected by dust extinction and additionally NIR luminosities provide reliable standard candles. To rule out the random influences of diverse SNe, a large number of light curves were employed. This choice favoured those light curves in which dust absorption is not very relevant. In fact, the dust extinction is the main uncertainty that reduces the precision of the optical light curves. A hierarchical Bayesian analysis was applied to retrieve NIR light curves of SNe Ia, the distribution of the intrinsic fluctuation of the light curves and the dust distribution. This Bayesian method was employed to evaluate the as-yet not explored SN parameters. Through a non-parametric method, they locally shaped the light curves assuming local parameters. The lack of data can also be managed using this method, deducing them in the case of incomplete time coverage or variability in the light curves. A hierarchical Bayesian method was employed by Mandel *et al* (2011) to properly describe randomized effects on the redshift, such as dust extinction, intrinsic fluctuations of the light curves and distances. Inferential models were also developed for optical and NIR ranges. Applying a Markov chain Monte Carlo algorithm re-sampling the global posterior probability density, this work was an update of that by Mandel *et al* (2009). The advantage of employing both NIR and optical samples is dual: the dust extinction is attenuated and the estimation of the SN Ia distances is more accurate by 60%. Indeed, through cross-validation, for SNe with optical and NIR data, the error in $\mu(z)$ was 0.11 mag in comparison to 0.15 mag for optical SNe only. For this reason, to use these objects as distance indicators, further investigation of SNe Ia in the NIR range was strongly advisable.

Unlike this method, for cosmological studies of GRBs, a Bayesian analysis was applied (Reichart *et al* 1999, D'Agostini 2005). It was stressed that this method is not hierarchical. In this way the intrinsic scatter (σ_{int}) was characterized by global, not local, parameters. Randomness, errors, intrinsic SN fluctuation, dust extinction, reddening, velocities and distances for each SN were included in this parametric approach. The main quantities of the SN Ia light curves were: intrinsic fluctuation, colour and luminosity in several wavelengths. The distribution of host galaxy dust was dominated by the dust parameters. The approach associated the dust parameters randomly to every SN light curve from the distribution of the real parameters. To produce a light curve with noise similar to the observed multi-wavelength light curve, the dust parameters were combined with $\mu(z)$ and the light curves. For the near universe $\mu(z)$ is a function not only of the redshift through the Hubble law, but also of a noise component indicating random velocities of the host galaxies. This random procedure was applied to each SN of the sample. To tighten the constraints of the cosmological parameters for SN Ia data from the SALT-II light curve fitter, March *et al* (2011) employed a Bayesian hierarchical method. Diminishing biases by 2–3 times through a simulated sample, their analysis reduced the confidence range of the cosmological parameters. Furthermore, they recovered a posterior probability distribution for the scatter of the intrinsic magnitude of SNe. Through this procedure, for the up-to-date SNe Ia and BAO and CMB samples, $\Omega_M = 0.28 \pm 0.02$ and $\Omega_\Lambda = 0.73 \pm 0.01$ were found for $w = -1$, and $\Omega_M = 0.28 \pm 0.01$ and $w = -0.90 \pm 0.05$ were found for the flat model. Finally, σ_{int} of the B magnitude of SN Ia data was constrained providing $\sigma_\mu = 0.13 \pm 0.01$ mag.

Employing a BAyesian HierArchical Modeling (the BAHAMAS software) for the analysis of supernova cosmology, 740 SNe Ia from the sample of the 'Joint

Light-curve Analysis' (JLA) were investigated by Shariff *et al* (2016). Combining the JLA and Planck CMB samples, the cosmological parameters were evaluated in comparison to the standard analysis. They pointed out significant differences in these quantities: $\Omega_M = 0.399 \pm 0.027$ and $w = -0.910 \pm 0.045$, 2.8σ and 1.6σ larger than earlier results, respectively. A question naturally arose: is this difference due to two subsamples with distinct host galaxy masses or to a 4σ discrepancy between the colour corrections at $z = 0.662$? It was concluded that neither of these was the reason for the difference in the results. Therefore, this is still an unsolved question.

Furthermore, the issues in SN cosmological methods regarding the outlier treatment, selection biases, diverse observations, σ_{int}, shape and colour standardization of the correlations were investigated by Rubin *et al* (2015). Employing a brand new Bayesian model, UNITY (Unified Nonlinear Inference for Type-Ia cosmologY), to a number of observed SN light curves, they improved the outcomes, achieving smaller statistical and systematic errors. They confirmed that SNe Ia demand nonlinear shape and colour standardizations, but these nonlinear correlations were taken into account in a proper statistical way in this procedure. In this analysis, the validation of the procedure was first made on a simulated sample, and then reiterated with a real sample. They did not find any change in the results.

Considering the larger up-to-date sample of SNe Ia, Nielsen *et al* (2016) examined the standard cosmological model to check if a larger sample can reproduce the same results obtained with the SN sample employed for the discovery of the cosmic acceleration. Employing the corrections to the absolute magnitudes to describe the fluctuation of the light curve and the dust extinction, the rate of expansion was found to be almost constant. This surprising result provided only negligible proof for the cosmic acceleration.

From all these analyses, GRBs, observed at higher redshift than SNe Ia, may explain the differences in the cosmological parameters claimed by March *et al* (2011). In addition, the study of GRBs has the potential to cast light on cosmic acceleration, as investigated in Nielsen *et al* (2016). To have a precision cosmology in agreement with that for SNe Ia, the strategy for the study of GRBs for cosmological purposes should be enhanced. This is the main goal of forthcoming space missions such as *SVOM* (Cordier *et al* 2015) and *THESEUS* (Amati *et al* 2017).

Finally, analogies and differences in the methodology adopted to use GRBs and SNe Ia as cosmological tools are described. The circularity problem affecting GRBs (see section 5.2.1) can be eluded through parameters independent of the cosmological model. Instead, the Cepheids can properly calibrate SNe Ia in the same galaxy, especially those at low z. Therefore, the calibration is not a relevant problem for SNe Ia. Furthermore, unlike the non-parametric method by Mandel *et al* (2011), the procedure for GRBs is parametric and employed through established correlations, not through individual light curves. Indeed, in the non-parametric procedure, the light curve first has to be reproduced locally to represent its behaviour in every phase and energy range. Only after that is it possible to obtain the description of the whole light curve in all the energy ranges. A method to treat a lack of data is also given by the probabilistic hierarchical approach. However, for GRBs there is still no such elaborated procedure for analysing correlations among prompt parameters. For this reason, selection effects can be present. Nevertheless, tighter cosmological parameters could be provided by the local description through the application of this procedure to GRBs. The light curves in the *Swift* repository are already unabsorbed, for details see Evans *et al* (2009). Thus, the dust extinction is not a variable of the hierarchical method. The GRB intrinsic scatter σ_{int} is evaluated parametrically. Furthermore, a training sample in the light curves has not yet been determined from the GRB investigation. To choose this training set and extend the approaches for SNe Ia to GRBs, the 'Gold' set of GRBs with flat plateau phases (Dainotti *et al* 2016) alone can supply a fundamental aid, without any additional GRB observational sample (Dainotti *et al* 2017).

References

Amati L 2012 Cosmology with the $E_{p,i}$–E_{iso} correlation of gamma-ray bursts *Int. J. Mod. Phys. Conf. Ser.* **12** 19–27

Amati L and Della Valle M 2013 Measuring cosmological parameters with gamma ray bursts *Int. J. Mod. Phys.* D **22** 1330028

Amati L, Guidorzi C, Frontera F, Della Valle M, Finelli F, Landi R and Montanari E 2008 Measuring the cosmological parameters with the $E_{p,i}$–E_{iso} correlation of gamma-ray bursts *Mon. Not. R. Astron. Soc.* **391** 577–84

Amati L, Frontera F and Guidorzi C 2017 Extremely energetic *Fermi* gamma-ray bursts obey spectral energy correlations *Astron. Astrophys.* **508** 173

Andreon S 2013 Understanding better (some) astronomical data using Bayesian methods *Astrostatistical Challenges for the New Astronomy* ed J M Hilbe (Berlin: Springer) pp 41–62

Atteia J-L 2003 A simple empirical redshift indicator for gamma-ray bursts *Astron. Astrophys.* **407** L1–4

Burgess J M 2015 The rest-frame Golenetskii correlation via a hierarchical Bayesian analysis, arXiv: 1512.01059

Cano Z 2014 Gamma-ray burst supernovae as standardizable candles *Astrophys. J.* **794** 121

Capozziello S and Izzo L 2008 Cosmography by gamma ray bursts *Astron. Astrophys.* **490** 31–6

Cordier B *et al* 2015 The SVOM gamma-ray burst mission, *Proc. of the Conf. Swift: 10 Years of Discovery (Rome, Italy 2–5 December 2014)* 1–13

D'Agostini G 2005 Fits, and especially linear fits, with errors on both axes, extra variance of the data points and other complications, arXiv: physics/0511182

Dai Z G, Liang E W and Xu D 2004 Constraining Ω_M and dark energy with gamma-ray bursts *Astrophys. J.* **612** L101–4

Dainotti M G, Cardone V F and Capozziello S 2008 A time–luminosity correlation for γ-ray bursts in the x-rays *Mon. Not. R. Astron. Soc.* **391** L79–83

Dainotti M G, Fabrizio Cardone V, Capozziello S, Ostrowski M and Willingale R 2011 Study of possible systematics in the L^*_X–T^*_a correlation of gamma-ray bursts *Astrophys. J.* **730** 135

Dainotti M G, Postnikov S, Hernandez X and Ostroski M 2016 A fundamental plane for long gamma-ray bursts with X-ray plateaus *Astrophys. J. Lett.* **825** L20

Dainotti M G, Nagataki S, Maeda K, Postnikov S and Pian E 2017 A study of gamma ray bursts with afterglow plateau phases associated with supernovae *Astron. Astrophys.* **600** A98

Demianski M, Piedipalumbo E, Sawant D and Amati L 2017 Cosmology with gamma-ray bursts. I. The Hubble diagram through the calibrated $E_{p,\Gamma}$–E_{iso} correlation *Astron. Astrophys.* **598** A112

Evans P A *et al* 2009 Methods and results of an automatic analysis of a complete sample of *Swift*-XRT observations of GRBs *Mon. Not. R. Astron. Soc.* **397** 1177–201

Firmani C, Ghisellini G, Ghirlanda G and Avila-Reese V 2005 A new method optimized to use gamma-ray bursts as cosmic rulers *Mon. Not. R. Astron. Soc.* **360** L1–5

Ghirlanda G 2009 Advances on GRB as cosmological tools *AIP Conf. Ser.* **111** 579–86

Ghirlanda G, Ghisellini G and Lazzati D 2004 The collimation-corrected gamma-ray burst energies correlate with the peak energy of their νF_ν spectrum *Astrophys. J.* **616** 331–8

Ghirlanda G, Ghisellini G and Firmani C 2006 Gamma-ray bursts as standard candles to constrain the cosmological parameters *New J. Phys.* **8** 123

Ghirlanda G, Nava L, Ghisellini G, Celotti A and Firmani C 2009 Short versus long gamma-ray bursts: spectra, energetics, and luminosities *Astron. Astrophys.* **496** 585–95

Guiriec S *et al* 2013 Evidence for a photospheric component in the prompt emission of the short GRB 120323A and its effects on the GRB hardness–luminosity relation *Astrophys. J.* **770** 32

Guiriec S, Gonzalez M M, Sacahui J R, Kouveliotou C, Gehrels N and McEnery J 2016 *CGRO*/BATSE data support the new paradigm for GRB prompt emission and the new L_i^{nTh}–$E_{\mathrm{peak},i}^{\mathrm{nTh,rest}}$ relation *Astrophys. J.* **819** 79

Guiriec S *et al* 2016 A unified model for GRB prompt emission from optical to γ-rays: exploring GRBs as standard candles *Astrophys. J.* **831** 1

Kodama Y, Yonetoku D, Murakami T, Tanabe S, Tsutsui R and Nakamura T 2008 Gamma-ray bursts in $1.8 < z < 5.6$ suggest that the time variation of the dark energy is small *Mon. Not. R. Astron. Soc.* **391** L1–4

Li X and Hjorth J 2014 Light curve properties of supernovae associated with gamma-ray bursts, arXiv: 1407.3506

Liang E and Zhang B 2006 Calibration of gamma-ray burst luminosity indicators *Mon. Not. R. Astron. Soc.* **369** L37–41

Liang N, Xiao W K, Liu Y and Zhang S N 2008 A cosmology-independent calibration of gamma-ray burst luminosity relations and the Hubble diagram *Astrophys. J.* **685** 354–60

Lin H-N, Li X and Chang Z 2016 Model-independent distance calibration of high-redshift gamma-ray bursts and constrain on the ΛCDM model *Mon. Not. R. Astron. Soc.* **455** 2131–8

Mandel A *et al* 2011 Type Ia supernova light curve inference: hierarchical models in the optical and near-infrared *Astrophys. J.* **731** 2

Mandel K S, Wood-Vasey W M, Friedman A S and Kirshner R P 2009 Type Ia supernova light-curve inference: hierarchical Bayesian analysis in the near-infrared *Astrophys. J.* **704** 629–51

March M C *et al* 2011 Improved constraints on cosmological parameters from type Ia supernova data *Mon. Not. R. Astron. Soc.* **418** 4

Nielsen D *et al* 2016 Marginal evidence for cosmic acceleration from type Ia supernovae *Astrophys. J.* **35596** 6

Perlmutter S 2003 Supernovae, dark energy, and the accelerating universe *Phys. Today* **56** 53–62

Phillips M M 1993 The absolute magnitudes of type IA supernovae *Astrophys. J.* **413** L105–8

Planck Collaboration *et al* 2016 Planck 2015 results. XIII. Cosmological parameters *Astron. Astrophys.* **594** A13

Qi S and Lu T 2010 A new luminosity relation for gamma-ray bursts and its implication *Astrophys. J.* **717** 1274–8

Reichart D E *et al* 1999 A Bayesian inference analysis of the x-ray cluster luminosity–temperature relation *Astrophys. J.* **516** 1

Riess A G *et al* 2004 Type Ia supernova discoveries at $z > 1$ from the *Hubble Space Telescope*: evidence for past deceleration and constraints on dark energy evolution *Astrophys. J.* **607** 665–87

Rubin D *et al* 2015 UNITY: confronting supernova cosmology's statistical and systematic uncertainties in a unified Bayesian framework *Astrophys. J.* **813** 2

Shariff H *et al* 2016 BAHAMAS: new SN Ia analysis reveals inconsistencies with standard cosmology *Astrophys. J.* **827** 1

Tsutsui R, Nakamura T, Yonetoku D, Murakami T, Kodama Y and Takahashi K 2009a Cosmological constraints from calibrated Yonetoku and Amati relation suggest fundamental plane of gamma-ray bursts *J. Cosmol. Astropart. Phys.* **8** 015

Tsutsui R, Nakamura T, Yonetoku D, Murakami T, Tanabe S, Kodama Y and Takahashi K 2009b Constraints on w_0 and w_a of dark energy from high-redshift gamma-ray bursts *Mon. Not. R. Astron. Soc.* **394** L31–5

Tsutsui R, Nakamura T, Yonetoku D, Murakami T and Takahashi K 2010 Intrisic dispersion of correlations among E_p, L_p, and E_{iso} of gamma ray bursts depends on the quality of data set, arXiv: 1012.3009

Tsutsui R, Yonetoku D, Nakamura T, Takahashi K and Morihara Y 2013 Possible existence of the E_p–L_p and E_p–E_{iso} correlations for short gamma-ray bursts with a factor 5–100 dimmer than those for long gamma-ray bursts *Mon. Not. R. Astron. Soc.* **431** 1398–404

Vitagliano V, Xia J-Q, Liberati S and Viel M 2010 High-redshift cosmography *J. Cosmol. Astropart. Phys.* **3** 005

Wang F Y and Dai Z G 2009 High-redshift star formation rate up to $z = 8.3$ derived from gamma-ray bursts and influence of background cosmology *Mon. Not. R. Astron. Soc.* **400** L10–4

Wang F Y, Dai Z G and Qi S 2009 Probing the cosmographic parameters to distinguish between dark energy and modified gravity models *Astron. Astrophys.* **507** 53–9

Wang F-Y, Qi S and Dai Z-G 2011 The updated luminosity correlations of gamma-ray bursts and cosmological implications *Mon. Not. R. Astron. Soc.* **415** 3423–33

Wang F Y, Dai Z G and Liang E W 2015 Gamma-ray burst cosmology *New Astron. Rev.* **67** 1–17

Wei H and Zhang S-N 2009 Reconstructing the cosmic expansion history up to redshift $z = 6.29$ with the calibrated gamma-ray bursts *Eur. Phys. J.* C **63** 139–47

Xu D, Dai Z G and Liang E W 2005 Can gamma-ray bursts be used to measure cosmology? A further analysis *Astrophys. J.* **633** 603–10

Yonetoku D, Murakami T, Nakamura T, Yamazaki R, Inoue A K and Ioka K 2004 Gamma-ray burst formation rate inferred from the spectral peak energy–peak luminosity relation *Astrophys. J.* **609** 935–51

Chapter 6

The afterglow relations

6.1 The correlations between afterglow parameters

The main correlations between physical parameters of the afterglow phase presented in the literature are: the LT correlation (Dainotti *et al* 2008), the unified $L_{X,a}-T^*_{X,a}$ and $L_{O,a}-T^*_{O,a}$ correlations (Ghisellini *et al* 2009), and the $L_{O,200s}-\alpha_{O,>200s}$ correlation (Oates *et al* 2012).

6.1.1 The LT correlation ($L_{X,a}-T^*_{X,a}$)

The LT correlation was the first to investigate the plateau phase features. This is an anti-correlation between $L_{X,a}$ and $T^*_{X,a}$, for definitions see section 3.2.

It was discovered by Dainotti *et al* (2008) employing 33 LGRBs detected by XRT. This sample is derived from the one analysed by W07 by selecting GRBs with known z and reliable spectral parameters. The LT correlation reads as follows:

$$\log L_{X,a} = a + b \times \log T^*_{X,a}, \tag{6.1}$$

with $\log L_{X,a}$ in erg s^{-1} and $T^*_{X,a}$ in seconds, a normalization $a = 48.54$, a slope $b = -0.74^{+0.20}_{-0.19}$ and $\sigma_{int} = 0.43$, and a Spearman correlation coefficient $\rho = -0.74$. $L_{X,a}$ has been calculated in the *Swift* XRT energy band using the formula:

$$L_{X,a}(z) = 4\pi D^2_L(z, \Omega_M, h)F_{X,a} \times K \tag{6.2}$$

where $D_L(z, \Omega_M, h)$ is the GRB luminosity distance for a given z, $F_{X,a}$ the flux in the X-ray energy band at the end of the plateau phase and $K = \frac{1}{(1+z)(1-\beta_{X,a})}$ the K-correction for cosmic expansion (Bloom *et al* 2001). From this anti-correlation it is suggested that the shorter the plateau phase duration, the brighter the plateau phase. Since the errors on both the variables $\log T_{X,a}$ and $\log L_{X,a}$ are comparable, it is more appropriate to employ a method that takes into account both the error bars rather than the simple method of the Levenberg–Marquardt algorithm. The latter

doi:10.1088/2053-2563/aae15cch6

method considers that the errors in one variable are negligible compared to the errors in the other variable. Thus, the fitting method by D'Agostini (2005), which relies on Bayesian assumptions and also accounts for σ_{int} of an unknown nature, have to be employed. However, Dainotti *et al* (2008) checked that the results of both the D'Agostini method and the Levenberg–Marquardt algorithm are in agreement within 1σ. Clearly, if the final goal is to employ this correlation as a model discriminator and as a cosmological tool, it is mandatory to reduce as much as possible the scatter of this correlation. To this end, a subset of GRBs taken from the initial sample of 33 GRBs has been chosen with definite selection criteria: $\log L_{X,a} > 45$ and $1 \leqslant \log T^*_{X,a} \leqslant 5$. With this choice the sample dropped to 28 GRBs obtaining a correlation with a much smaller scatter (23%) and with $(a, b, \sigma_{\text{int}}) = (48.09, -0.58 \pm 0.18, 0.33)$.

From the investigation of 33 LGRBs during the late prompt phase in optical and X-ray bands, Ghisellini *et al* (2009) confirmed the value of the slope for the LT correlation, $b = -0.58^{+0.18}_{-0.18}$, when this particular selection choice is applied: $1 \leqslant \log T^*_{X,a} \leqslant 5$.

Afterwards, a slope $b = -1.06^{+0.27}_{-0.28}$ was recovered by Dainotti *et al* (2010), considering 62 LGRBs. When taking into account only the eight Intermediate class (IC) GRBs in the sample $b = -1.72^{+0.22}_{-0.21}$ was obtained. Additionally, introducing the quantity $\sigma_E = (\sigma^2_{\log L_{X,a}} + \sigma^2_{\log T^*_{X,a}})^{1/2}$, with $\sigma_{\log L_{X,a}}$ and $\sigma_{\log T^*_{X,a}}$, the errors in luminosity and time respectively, Dainotti *et al* (2010) claimed that for the eight GRBs with the smallest errors ($\sigma_E < 0.095$) the LT correlation has a slope $b = -1.05^{+0.19}_{-0.20}$ (see the left panel of figure 6.1, the right panel of figure 6.2, and table 6.1). Likewise, Bernardini *et al* (2012) and Sultana *et al* (2012), from the analysis of 64 and 14 LGRBs, respectively, retrieved a slope $b \approx -1$ (for details see table 6.1).

Later, Dainotti *et al* (2011) found a correlation with $b = -1.20^{+0.27}_{-0.30}$, using an enlarged dataset of 77 LGRBs. Instead, Mangano *et al* (2012) obtained a steeper

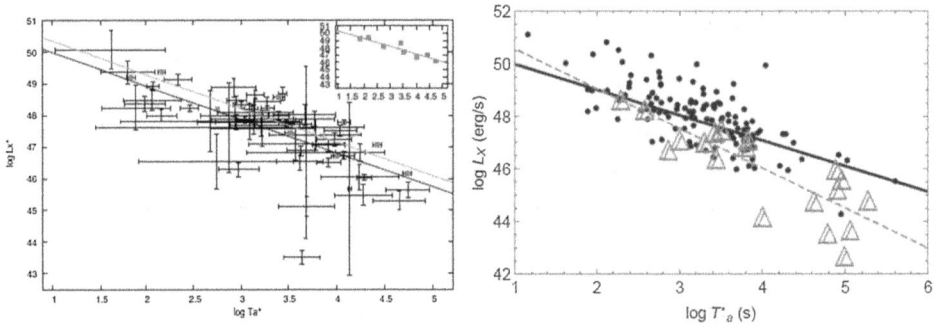

Figure 6.1. Left panel: the $\log L_{X,a} = \log L^*_X - \log T^*_{X,a}$ distribution for 62 LGRBs with $\sigma_E < 4$ and the best-fit indicated by the black line. The eight red points with the lowest errors in the sample are fitted with a red line and are also displayed in the subpanel in the upper right-hand corner. (Reproduced with permission from Dainotti *et al* (2010). Copyright 2010 The American Astronomical Society.) Right panel: the blue points (fitted with a solid line of the same colour) indicate 128 GRBs of the LONG-NO-SNe set, while the red empty triangles (fitted with a dashed line of the same colour) represent 19 GRBs of the LONG-SNe set (Reproduced with permission from Dainotti *et al* (2017). Copyright 2017 ESO.)

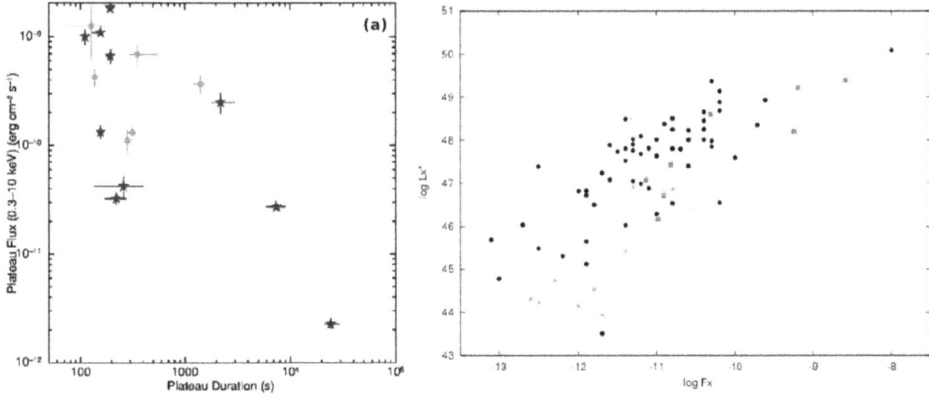

Figure 6.2. Left panel: flux versus time for the plateau phase of 22 SGRBs. GRB light curves with two or more breaks are shown as blue stars, while GRB light curves with only one break are represented by green circles (Reproduced from Rowlinson *et al* (2013). By permission of Oxford University Press on behalf of the Royal Astronomical Society.) Right panel: the log $L_{X,a}$–log $L_{F,a}$ distribution for the whole GRB set. Red squares represent the eight GRBs with the lowest errors in the sample, while the green triangles show the IC GRBs. (Reproduced with permission from Dainotti *et al* (2010). Copyright 2010 The American Astronomical Scociety.)

slope ($b = -1.38^{+0.16}_{-0.16}$), but mostly due to the fact that they used some GRBs with no evident plateau phase in their set of 50 LGRBs. They fitted this sample with a broken power-law as a fitting function instead of the W07 function. From the outcomes of all these works, it was concluded that when the number of GRBs is enlarged a steepening of the slope is pointed out.

For this reason, Dainotti *et al* (2013a) studied the selection biases affecting the correlation. They found that, even if the steepening of the correlation slope is caused by selection biases, the intrinsic (namely, after it has been fully corrected for selection biases) slope (b_{int}) of the correlation can be computed through a robust statistical analysis. Eventually it has value $b_{int} = -1.07^{+0.09}_{-0.14}$, (see section 8). In conclusion, Dainotti *et al* (2013a) and Rowlinson *et al* (2014) corroborated the outcomes from Dainotti *et al* (2010) by using enlarged samples of 101 and 159 GRBs, respectively.

The previous results by Dainotti *et al* (2013a) were validated from Dainotti *et al* (2015) using a wider dataset of 123 LGRBs detected by *Swift*/XRT.

To diminish the scatter of the LT correlation, using 176 *Swift* GRBs Del Vecchio *et al* (2016) studied the temporal decay indices after the plateau phase, $\alpha_{X,a}$, obtained through two different fitting functions: a simple power-law, taking into account the declining part after the plateau phase, and the W07 one (see section 1.2). Given that the findings do not depend on the chosen function, Del Vecchio *et al* (2016) looked for some common characteristics in GRBs to use them as standard candles and to constrain the physical models presented by Hascoët *et al* (2014) for the plateau phase emission. As a result, it was pointed out that the $\alpha_{X,a}$ parameter can give rise to different LT correlations for low and high luminosity GRBs, indicating a difference in the density of the external medium.

Table 6.1. Summary of the LT correlation. All GRBs are observed by *Swift*. The authors are indicated in the first column, the size of the dataset is displayed in the second and the type of GRB (S = short, l = Long, IC = intermediate) is listed in the third. The slope and normalization of the correlation are shown in the fourth and the fifth columns, respectively. The correlation coefficient and the chance probability, P, are presented in the last two columns, respectively. (Table from Dainotti and Del Vecchio (2017).)

Author	N	Type	Slope	Norm	Corr. coeff.	P
Dainotti *et al* (2008)	28	$1 < T^*_{X,a} < 5$	$-0.58^{+0.18}_{-0.18}$	48.09	-0.80	1.6×10^{-7}
Dainotti *et al* (2008)	33	All GRBs	$-0.74^{+0.20}_{-0.19}$	48.54	-0.74	10^{-9}
Cardone *et al* (2009)	28	L	$-0.58^{+0.18}_{-0.18}$	48.09	-0.74	10^{-9}
Ghisellini *et al* (2009)	33	L	$-0.58^{+0.18}_{-0.18}$	48.09	-0.74	10^{-9}
Cardone *et al* (2010)	66	L	$-1.04^{+0.23}_{-0.22}$	$50.22^{+0.77}_{-0.76}$	-0.68	7.6×10^{-9}
Dainotti *et al* (2010)	62	L	$-1.06^{+0.27}_{-0.28}$	$51.06^{+1.02}_{-1.02}$	-0.76	1.85×10^{-11}
Dainotti *et al* (2010)	8	High luminosity	$-1.05^{+0.19}_{-0.20}$	$51.39^{+0.90}_{-0.90}$	-0.93	1.7×10^{-2}
Dainotti *et al* (2010)	8	IC	$-1.72^{+0.22}_{-0.21}$	$52.57^{+1.04}_{-1.04}$	-0.66	7.4×10^{-2}
Dainotti *et al* (2011a)	77	L	$-1.20^{+0.27}_{-0.30}$	$51.04^{+0.27}_{-0.30}$	-0.69	7.7×10^{-8}
Sultana *et al* (2012)	14	L	$-1.10^{+0.03}_{-0.03}$	$51.57^{+0.10}_{-0.10}$	-0.88	10^{-5}
Bernardini *et al* (2012)	64	L	$-1.06^{+0.06}_{-0.06}$	51.06	-0.68	7.6×10^{-9}
Mangano *et al* (2012)	50	L	$-1.38^{+0.16}_{-0.16}$	$52.2^{+0.06}_{-0.06}$	-0.81	2.4×10^{-10}
Dainotti *et al* (2013a)	101	Simulated	$-1.52^{+0.04}_{-0.24}$	$53.27^{+0.54}_{-0.48}$	-0.74	10^{-18}
Dainotti *et al* (2013b)	101	ALL intrinsic	$-1.07^{+0.09}_{-0.14}$	52.94	-0.74	10^{-18}
Dainotti *et al* (2013b)	101	All GRBs	$-1.32^{+0.18}_{-0.17}$	$52.8^{+0.9}_{-0.3}$	-0.74	10^{-18}
Dainotti *et al* (2013b)	101	Without short	$-1.27^{+0.18}_{-0.26}$	52.94	-0.74	10^{-18}
Postnikov *et al* (2014)	101	L ($z < 1.4$)	$-1.51^{+0.26}_{-0.27}$	$53.27^{+0.54}_{-0.48}$	-0.74	10^{-18}
Rowlinson *et al* (2014)	159	Intrinsic	$-1.07^{+0.09}_{-0.14}$	52.94	-0.74	10^{-18}
Rowlinson *et al* (2014)	159	Observed	$-1.40^{+0.19}_{-0.19}$	$52.73^{+0.52}_{-0.52}$	-0.74	10^{-18}
Rowlinson *et al* (2014)	159	simulated	$-1.30^{+0.03}_{-0.03}$	$52.73^{+0.52}_{-0.52}$	-0.74	10^{-18}
Dainotti *et al* (2015)	123	L	$-0.90^{+0.19}_{-0.17}$	$51.14^{+0.58}_{-0.58}$	-0.74	10^{-15}
Dainotti *et al* (2017)	19	L-SNe	$-1.5^{+0.3}_{-0.3}$	$51.85^{+0.94}_{-0.94}$	-0.83	5×10^{-6}

Recently, using a dataset of 176 *Swift* GRBs with measured redshifts, Dainotti *et al* (2017) found out that the group of LGRBs connected with SNe (LONG-SNe) displays an LT correlation with a significant Spearman correlation coefficient, ρ (see the right panel of figure 6.1). Comparing the group of LONG GRBs associated with SNe to that of LGRBs with no associated SNe (hereafter LONG-NO-SNe, 128 GRBs), they employed the Efron and Petrosian (1992) method to investigate whether there is an intrinsic difference between the slopes of these two subsamples. To this end, the first step was to verify if a different redshift evolution pertained to the 128 LONG-NO-SNe sample compared to the evolution associated with the previously analysed sample of 101 GRBs. In the latter sample all the GRB categories were considered together (Dainotti *et al* 2013a). This step is mandatory because the two subsamples are observed at different redshifts, namely the

LONG-NO-SNe sample is observed at higher z than the LONG-SNe sample. The second step is to test if the intrinsic slope, b_{int}, of the 128 LONG-NO-SNe sample is different from the observed slope of the same sample. The third step is to compare b_{int} of the two samples assuming the sample of the GRB-SNe is not affected by redshift evolution since it is observed at small redshifts. Through this analysis, they concluded that b_{int} for the LONG-NO-SNe group is the same as the observed one, thus no steepening of the slope for this sample is found. Therefore, they compared the slopes of the two samples, finding a statistical difference ($P = 0.005$) between the LONG-SNe sample with firm spectroscopic evidence and the LONG-NO-SNe sample. Since the observed slope of the LONG-SNe sample with firm spectroscopic evidence is -1.9, this value very probably implies that this sample does not demand a typical energy reservoir for the plateau phase. As a result, this study might envision a new path for future theoretical modelling. In this work it was also investigated how θ_{jet} can affect these results. If the same beaming angle of $10°$ is chosen for the whole LONG-NO-SNe sample, the slope of the sample does not change, but it experiences only a shift in the normalization. However, the beaming angle is larger for low luminous GRBs. Thus, assuming the beaming correction, the discrepancies between the two groups are only statistically significant at 10%. Therefore, this evidence opens two possible scenarios: the differences in the slopes may be due to the poor accuracy in the determination of θ_{jet} or to the presence of dim GRBs in the LONG-SNe sample which have not been previously corrected for the beaming angles. For these reasons, the discussion is not concluded, and thus it will be very helpful to collect new data for further analysis.

In table 6.1, the parameters a and b with ρ and P for the LT correlation are displayed.

Physical interpretation of the ($L_{X,a}$–$T^*_{X,a}$) correlation
In this section the theoretical interpretation of the LT correlation is presented. The accretion (Cannizzo and Gehrels 2009, Cannizzo *et al* 2011) and the magnetar models (Zhang and Mészáros 2001, Dall'Osso *et al* 2011, Rowlinson and O'Brien 2012, Rowlinson *et al* 2013, 2014, Stratta *et al* 2018) are the main theoretical descriptions for the LT correlation. In the first model the material moving around the GRB progenitor star and collapsing towards the core of the progenitor generates an accretion disk. After the material around the disk is squeezed by the gravitational forces, the GRB is emitted. Kumar *et al* (2008) suggested that in LGRBs the early steep decay phase slope can give constraints on the distribution of the radial density in the progenitor.

A steeper correlation slope ($-3/2$) than the observed one (~-1) was obtained by Cannizzo and Gehrels (2009), nevertheless it was comparable with the one obtained from the model by Yamazaki (2009). With a set of 62 LGRBs and few SGRBs Cannizzo *et al* (2011) analysed the simulated fall-back disks around the BH. They pointed out that a circularization radius with a value of 10^{10}–10^{11} cm provides a value for the plateau phase duration of 10^4 s for LGRBs, with the starting fall-back mass $10^{-4}M_\odot$, see the left panel of figure 6.3). In the case of SGRBs the radius is

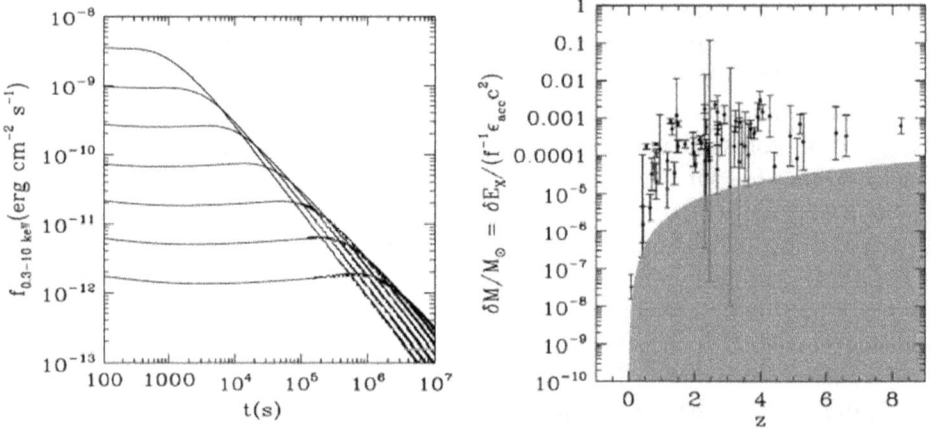

Figure 6.3. Left panel: light curves for LGRBs obtained through the constant ($10^{-4}M_\odot$) fall-back disk mass but modifying the normalization and initial radius. Right panel: the total accretion mass for the plateau added to the later decay phases of GRBs for 62 LGRBs from Dainotti *et al* (2010). A limiting XRT detection flux level of 10^{-12} erg cm^{-1} s^{-1} with a plateau phase lasting 10^4 s is displayed by the red area. A value of the beaming factor $f = 1/300$ and a value of the net efficiency fuelling the X-ray flux $\epsilon_{net} = \epsilon_{acc}\epsilon_X = 0.03$ are considered. (Reproduced with permission from Cannizzo *et al* (2011). Copyright 2011 The American Astronomical Society.)

evaluated to be around 10^8 cm. A lower limit for the accreting mass is provided by the LT correlation and has a value between 10^{-4} and $10^{-3}M_\odot$[1]. Therefore, it was concluded that if a standard energy reservoir for the fall-back mass is taken into account, the LT correlation can be recovered (see the right panel of figure 6.3). However, in this work the very steep decay following the prompt phase is not examined, different to Lindner *et al* (2010).

Instead, with regard to the magnetar model, Zhang and Mészáros (2001) investigated the effects of a strongly magnetized millisecond pulsar on the GRB afterglow phase. They concluded that, for particular initial pulsar rotation period and magnetic field, the afterglow phase should display an achromatic bump lasting from minutes to months. This feature can determine some limits on the progenitor models. Later, Dall'Osso *et al* (2011) recovered the GRB shallow decay phase characteristics by fitting a few *Swift* XRT light curves and computing the parameters for the magnetar model (i.e. a spin period of 1–3 ms and magnetic field ~10^{14}–10^{15} G). They also reproduced within this scenario the LT correlation (see the left panel of figure 6.4).

Similarly to Dall'Osso *et al* (2011), Bernardini *et al* (2012), analysing a dataset of 64 LGRBs, verified that the GRB light curve phase, displaying a shallow decay which obey the LT correlation, can be well described in the framework of the magnetar model. Rowlinson and O'Brien (2012) and Rowlinson *et al* (2013) claimed

---

[1] The following estimate can be calculated assuming the total accretion mass $\Delta M/M = \Delta E_X/f^{-1}*\epsilon_{acc}^*c^2$, with c the light speed, f the beaming factor, ϵ_{acc} the efficiency of accretion onto the BH and E_X the plateau energy plus later decaying stages.

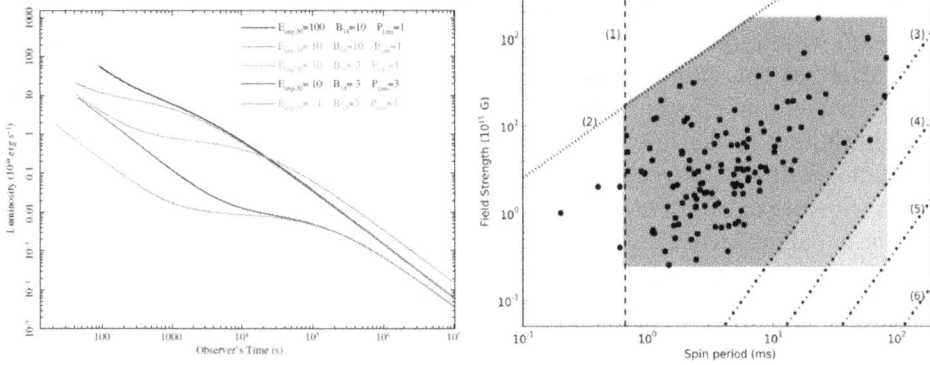

Figure 6.4. Left panel: five light curves computed through the variation of the afterglow energy and the NS spin period, P, at the beginning and the dipole magnetic field, B. (Reproduced with permission from Dall'Osso *et al* (2011). Copyright 2011 ESO.) Right panel: the regular distributions of B and P assumed in the simulation of the observed magnetar plateau phases are indicated by the grey shaded area. The magnetar model (Lyons *et al* 2010, Dall'Osso *et al* 2011, Bernardini *et al* 2012, Gompertz *et al* 2013, Rowlinson *et al* 2013, Yi *et al* 2014, Lü and Zhang 2014) was used for the fit of the light curves to recover upper and lower boundaries on B and P. For an explanation of the different areas of the plot see Rowlinson *et al* (2014). From these distributions it was concluded that the efficiencies and the beaming angles of the GRBs in their sample are small. (Reproduced from Rowlinson *et al* (2014). By permission of Oxford University Press on behalf of the Royal Astronomical Society.)

that the energy injection produced by a magnetar is able to nicely represent most SGRBs.

Additionally, by employing 159 *Swift* GRBs Rowlinson *et al* (2014) proved that the central engine model recovers the LT correlation. This result was obtained assuming that the compact object is emitting energy into the FS. The luminosity and plateau phase duration are given by:

$$\log L_{X,a} \sim \log(B_p^2 P_0^{-4} R^6) \tag{6.3}$$

and

$$\log T_{X,a}^* = \log(2.05 \times I B_p^{-2} P_0^2 R^{-6}), \tag{6.4}$$

where $T_{X,a}^*$ is in units of 10^3 s, $L_{X,a}$ is in units of 10^{49} erg s^{-1}, the moment of inertia in units of 10^{45} g cm^2 is expressed by I, the strength of the magnetic field at the poles in units of 10^{15} G is indicated by B_p, the NS radius in units of 10^6 cm is given by R and the initial period of the compact object in milliseconds is described by P_0. After placing the radius from equation (6.4) into equation (6.3), it was obtained that:

$$\log (L_{X,a}) \sim \log (10^{52} I^{-1} P_0^{-2}) - \log (T_{X,a}^*). \tag{6.5}$$

This indicates the existence of an intrinsic $\log L_{X,a} \sim -\log T_{X,a}^*$ correlation. Even if some magnetar plateau phases seem in disagreement with energy injection into the FS, Rowlinson *et al* (2014) pointed out that the rotational energy transfer from the compact object to the luminosity of the plateau phase has $\leqslant 20\%$ efficiency and this emission is narrowly beamed. Moreover, the intrinsic LT correlation slope is recovered within the framework of the spin-down of a recently produced magnetar at 1σ level (see the right panel of figure 6.4). The scatter in this correlation is caused by the different initial spin periods.

Contributing to the discussion of all these papers on the connection of the LT correlation with the magnetar model, Rea *et al* (2015) employed a sample of *Swift* GRBs with measured z observed from its launch until August 2014. They pointed out that the initial magnetic field distribution of GRB-magnetars is not in agreement with that of the galactic magnetars. For this reason, the GRB-magnetar model is correct only if two types of magnetar progenitors are allowed. In particular, GRB-magnetars should be regarded as supermagnetars (magnetars with a huge initial magnetic field) and separately from galactic magnetars. Furthermore, from their analysis they put a constraint on the number of stable magnetars generated by a GRB in the Milky Way in the past Myr (about $\leqslant 16$). However, the core collapse SN rate is 10% lower than the galactic magnetar rate, while the rate of GRBs is much smaller. Thus, thus galactic magnetars cannot explain the production of GRBs. In addition, the GRB magnetars; spin-down rates have to be very fast in comparison to the default spin-down rates measured for magnetars, and the low GRB rate would not easily allow the observation of these supermagnetars. In conclusion, it was claimed that the results from this paper are in agreement with those from previous ones.

The first analytical analysis of the GRB lightcurves corrected by selection biases and with a meaningful statistical sample has been presented in Stratta *et al* (2018). In this paper they fit a sample of GRB X-ray plateaus, interestingly yielding a distribution in the magnetic field versus spin period (B-P) diagram consistent with P7/6 expected from the well-established physics of the spin-up line minimum period for Galactic millisecond pulsars. The normalization of the relation we obtain perfectly matches spin-up line predictions for the expected masses (roughly 1 solar mass) and radii (10 km) of newly born magnetars, and mass accretion rates consistent with GRB expectations of the well-established physics of the spin-up line for accreting Galactic X-ray pulsars.

Short GRBs with extended emission (SEE) appear towards the high period end of the distribution, while the long GRBs (LGRBs) appear towards the short period end. This result is consistent with spin-up limit expectations where the total accreted mass determines the position of the neutron star in the B-P diagram. The P-B distributions for LGRBs and SEE are statistically different, further supporting the idea that the fundamental plane relation, (Dainotti *et al* 2016, Dainotti *et al* 2017), is a powerful discriminant among those populations. Our conclusions are robust against suppositions regarding the GRB collimation angle and magnetar breaking index, which shifts the resulting magnetar properties parallel to the spin-up line, and strongly support a magnetar origin for GRBs presenting X-ray plateaus.

Still in the framework of the energy injection models, van Eerten (2014a) obtained a correlation between $F_{O,a} \sim T_{O,a}^{-0.78\pm0.08}$ (Panaitescu and Vestrand 2011, Li *et al* 2012), for definitions see section 3.2. The range of $E_{\gamma,\mathrm{iso}}$, the fraction of the magnetic energy, ϵ_B, and the initial density, n_0, generate the scatter in the correlation without

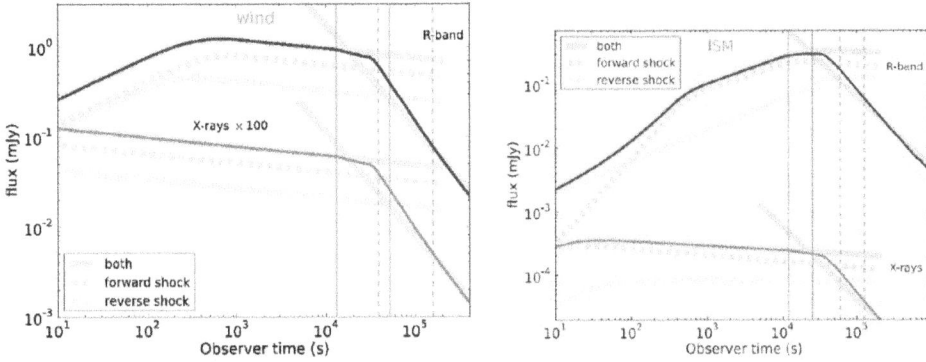

Figure 6.5. Light curves in the optical and X-ray ranges in the case of wind (left panel) and ISM (right panel) external media. The analytical curves for prolonged and impulsive energy injection are displayed by thick light grey curves. The area of FS radiation alone and that of RS radiation alone are represented by the thick dashed light grey and the thick dotted light grey functions, respectively. For additional details of the plots see van Eerten (2014a). (Reproduced from van Eerten (2014a). By permission of Oxford University Press on behalf of the Royal Astronomical Society.)

varying the slope. They concluded that both the medium characterized by the wind profile ($\propto A/r^2$, where A is a constant) and the ISM recovered the observed correlation within both the RS and FS models (see figure 6.5).

Alternatively, Sultana *et al* (2013) investigated the evolution of Γ during the whole light curves in the framework of the supercritical pile model. As a result, they obtained the LT correlation. The supercritical pile model supplies a description for the steep-decline and the plateau phase or the steep-decline and the power-law decay of the afterglow phase. Their main outcome was that the plateau phase in the evolution of Γ lasts less as the value of M_0c^2 diminishes, with M_0 indicating the initial rest mass of the flow. For this reason, the brighter the plateau phase, the shorter its duration. This implies a tinier value of M_0c^2, namely the energy.

In the framework of the thick shell model, from the analysis of the synchrotron emission Leventis *et al* (2014) claimed that this radiation is in agreement with the existence of the plateau phase (see the left panel of figure 6.6). From the study of the $\log F_{X,a}$–$\log T_{X,a}$ correlation they concluded that the energy emission through the RS is preferred to that through the FS (see the right panel of figure 6.6).

From the simulation of a GRB dataset by van Eerten (2014b), it was obtained that the observed LT correlation rejects thin shell models, but not basic thick ones. For thin shell models, the pre-deceleration radiation coming from a slow-moving component in a two-component or jet-type model leads to the plateau phase, but this description does not explain the observed LT correlation (see figure 6.7). On the other hand, this does not necessarily suggest that reliable fits are not feasible using a thin shell model. Therefore, additional investigation is mandatory to definitely rule out thin shell models. Instead, for the thick models, the late activity of the central engine, or further kinetic energy transfer from slower ejecta which interacts with the blast wave, generate the plateau phase. However, in this case it is difficult to

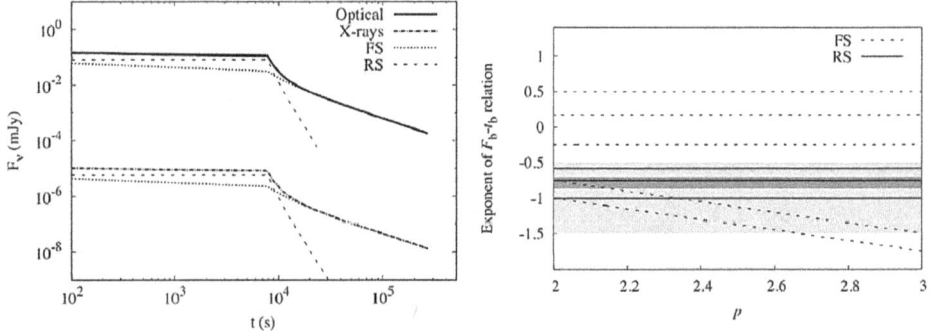

Figure 6.6. Left panel: light curves in optical and X-ray wavelengths before and after the injection break. The FS emission is indicated by a dotted line, while the RS one is displayed by the dashed line. The values of the variables assumed for the physical model are $E = 10^{51}$ erg, $n_1 = 50$ cm^{-3}, $\Delta t = 5 \times 10^3$ s, $\eta = 600$, q = 0, $\epsilon_e = \epsilon_B = 0.1$, $p = 2.3$, $\theta_j = 90°$, $d = 10^{28}$ cm and $z = 0.56$. Right panel: the $F_{X,a}$–$T_{X,a}$ correlation slope versus the electron distribution index for the FS and the RS. Results obtained by Panaitescu and Vestrand (2011) are presented by a lightly shaded area, while results by Li *et al* (2012) are displayed by a darker region. The five plausible indices for the FS are represented by the five dashed lines, while for the RS case the three plausible indices (not depending on the electron distribution index) are indicated by the three solid lines. (Reproduced from Leventis *et al* (2014). By permission of Oxford University Press on behalf of the Royal Astronomical Society.)

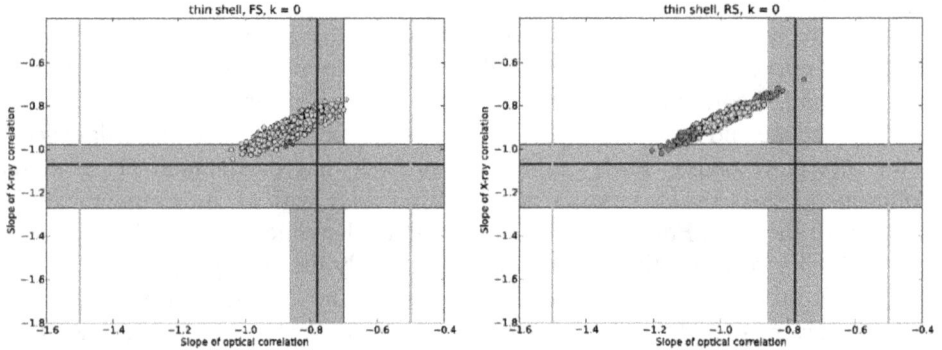

Figure 6.7. Simulated slopes of the X-ray (vertical axes) and optical (horizontal axes) LT correlations for 1000 thin shell samples considering the FS (left panel) or the RS (right panel) scenario. 1σ errors on both the optical and X-ray relations are represented by a grey region. Green dots and orange dots are compatible for both relations within 1σ and 3σ, respectively. The red dots are not consistent with either 1- or 3σ in either of the two relations. The errors in the optical LT correlation from Panaitescu and Vestrand (2011) are displayed by the vertical grey lines. (Reproduced from van Eerten (2014b). By permission of Oxford University Press on behalf of the Royal Astronomical Society.)

differentiate between FS and RS radiations, or homogeneous and stellar wind-type external media. Finally, another possibility would be the photospheric model from stratified jets (Ito *et al* 2014), which has not yet been examined on the LT correlation.

6.1.2 The unified $L_{X,a}-T_{X,a}^*$ and $L_{O,a}-T_{O,a}^*$ correlations

To analyse the X-ray and optical afterglow phases as a unique picture, it is mandatory to consider important characteristics of the optical luminosities. For this purpose, Boër and Gendre (2000), with a sample of eight GRBs, claimed that in the X-ray range the afterglow decay index ($\alpha_{X,a}$) distribution presents a bimodality, with the most luminous GRBs clustering around the value 1.6 and the faintest ones around the value 1.11. In the optical range, possibly due to host galaxy absorption, this behaviour is not visible.

Subsequently, Nardini *et al* (2006) pointed out the existence of a group of dark GRBs, namely GRBs with optical-to-X-ray spectral index, $\beta_{OX,a}$, smaller than 0.5, $\beta_{OX,a} \lesssim 0.5$. Indeed, the monochromatic $L_{O,12}$ (for definitions see section 3.2) of 24 LGRBs clustered at $\log L_{O,12} = 30.65$ erg s^{-1} Hz^{-1} with $\sigma_{int} = 0.28$. The dispersion of this distribution was smaller than that of $L_{X,12}$ and that of the ratio $L_{O,12}/E_{\gamma,\mathrm{prompt}}$, for definitions see section 3.2. From their analysis, three outliers were found with luminosity lower by a factor of ~ 15.

Liang and Zhang (2006a), with a sample of 44 GRBs, recovered a bimodal distribution of $L_{O,1d}$, for definitions see section 3.2, validating the previous results. Nardini *et al* (2008a) enlarged the dataset to 55 *Swift* LGRBs with measured z and rest-frame optical extinction, and investigated the selection biases in their observations. Their results were in agreement with previous works. Instead, no evidence of bimodality was found in the distributions of $L_{O,12}$, $L_{O,1d}$ and $L_{O,11}$ by Melandri *et al* (2008), Oates *et al* (2009), Zaninoni *et al* (2013) and Melandri *et al* (2014), who used 44, 24, 40 and 47 GRBs, respectively.

With the aim of looking for a common description of the GRB afterglow phase, Ghisellini *et al* (2009) analysed 33 *Swift* LGRBs in the X-ray range between 0.3 and 10 keV and in the optical R range (see the left and middle panels of figure 6.8). They employed the following function:

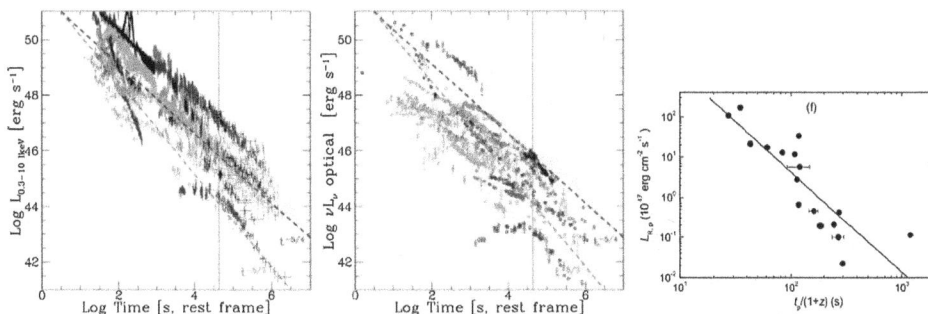

Figure 6.8. The X-rays (left panel) and optical (middle panel) light curves for the GRB dataset. $\log L_{X,12}$ and $\log L_{O,12}$ at a given rest-frame time are indicated by vertical lines, while the $\log t^{-5/4}$ (blue) and the $\log t^{-5/3}$ (red) trends are represented by the dashed lines. (Reproduced from Ghisellini *et al* (2009). By permission of Oxford University Press on behalf of the Royal Astronomical Society.) Right panel: the $T_{O,\mathrm{peak}}^*-L_{O,\mathrm{peak}}$ correlation with the best-fit line. In the picture, $L_{O,\mathrm{peak}}$ is equivalent to $L_{R,p}$. (Reproduced with permission from Liang *et al* (2010). Copyright 2010 The American Astronomical Society.)

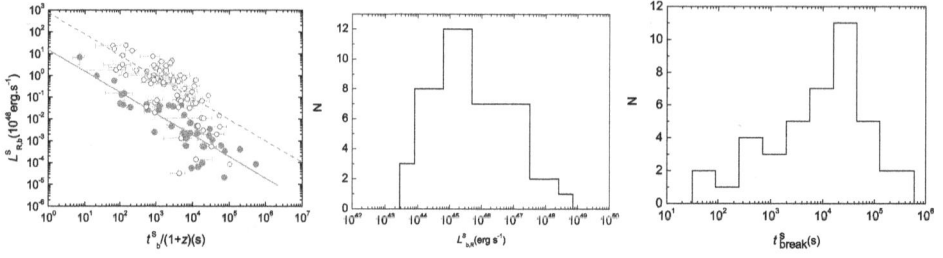

Figure 6.9. Left panel: the $L_{O,\,a}^{S} - T_{O,\,a}^{S,\,*}$ relation ($L_{R,p}^{S} - t_b$ in the picture) is represented by red circles, while the X-ray data from Dainotti *et al* (2010) are indicated by grey circles and the lines show the best-fit lines. Middle and right panels: the distributions of $L_{O,a}^{S}$ and $T_{O,a}^{S}$ for the whole GRB sample. (Reproduced with permission from Li *et al* (2012). Copyright 2012 The American Astronomical Society.)

$$L_L(\nu,\,t) = L_L(\nu,\,T_{X,a})\frac{(t/T_{X,t})^{-\alpha_{\nu,fl}}}{1 + (t/T_{X,t})^{\alpha_{\nu,st}-\alpha_{\nu,fl}}}. \qquad (6.6)$$

With this approximation they found good agreement with the GRB light curves, and they were able to reproduce the X-ray LT correlation.

Employing 32 *Swift* GRBs, Liang *et al* (2010) concluded that $L_{O,\text{peak}}$ in the R filter and $T_{O,\text{peak}}^{*}$ are anti-correlated (for definitions see section 3.2) with a slope $b = -2.49 \pm 0.39$ and $\rho = -0.90$ (see the right panel of figure 6.8). In conclusion, they claimed that a dimmer bump shows its maximum later than more luminous ones as well as a longer duration. Similarly, Panaitescu and Vestrand (2011) recovered a $\log F_{O,a} \sim \log T_{O,a}^{-1}$ anti-correlation with a sample of 37 *Swift* GRBs. They claimed the existence of a common process for the afterglow phase in the optical range, despite the large scatter in the optical band.

Later, Li *et al* (2012) found a correlation (see the left panel of figure 6.9) analogous to the LT one, but in the R energy range. They employed 39 GRBs with data in the optical range found in the literature. This correlation was between $L_{O,a}$, in units of 10^{48} erg s^{-1}, and $\log T_{O,a}^{*}$, for definitions see section 3.2. The quantities were defined for the shallow decay phase of the GRB light curves and denoted with the superscript S. They retrieved a slope $b = -0.78 \pm 0.08$, $\rho = 0.86$ and $P < 10^{-4}$.

$L_{O,a}^{S}$ fluctuated between 10^{43} and 10^{47} erg s^{-1}, and in several GRBs with an early break arrived at $\sim 10^{49}$ erg s^{-1} (see the middle panel of figure 6.9). $T_{O,a}^{S}$ varied between tens of seconds and days after the trigger of the GRB, with a peak at $T_{O,a}^{S} \sim 10^{4}$ s (see the right panel of figure 6.9). They concluded that optical light curves have similar behaviour to the X-ray light curves. Indeed, the correlation displayed in the left panel of figure 6.9, with a slope $b = -0.78 \pm 0.08$, reproduces the LT correlation for the X-ray flares. Table 6.2 summarizes b and P of the correlations mentioned in this section.

6.1.3 Physical interpretation of the unified $L_{X,a} - T_{X,a}^{*}$ and $L_{O,a} - T_{O,a}^{*}$ correlations

In the unified $L_{X,a} - T_{X,a}^{*}$ and $L_{O,a} - T_{O,a}^{*}$ correlations Ghisellini *et al* (2009) assumed the flux to be the sum of two components: one is the synchrotron radiation caused by

Table 6.2. Summary of the correlations presented in this section. The correlation in log scale is listed in the first column, the authors are presented in the second and the size of the dataset is displayed in the third. The correlation slope is indicated in the fourth column, while the correlation coefficient and the chance probability, P, are shown in the last two columns. (Table from Dainotti and Del Vecchio (2017).)

Correlations	Author	N	Slope	Corr. coeff.	P
$L_{O,\text{peak}} - T_{O,\text{peak}}$	Liang et al (2010)	32	-2.49 ± 0.39	-0.90	
$L_{O,a} - T_{O,a}$	Panaitescu and Vestrand (2011)	37	-1		
$L_{O,a}^{S} - T_{O,a}^{S}$	Li et al (2012)	39	-0.78 ± 0.08	0.86	10^{-4}

the standard FS, see section 2.2.1 for additional details, and the latter might be generated by a long-lived central engine responsible for the 'late prompt' emission. Although this treatment relies on an observational model, the achromatic and chromatic jet breaks in the data can be reliably described. Furthermore, they claimed that the decay slope of the late prompt phase ($\alpha_{X,a} = -5/4$) has a value analogous within the error bars to the estimate of the rate of accretion of the fall-back material ($\sim \log t^{-5/3}$). This situation is well represented by the blue and red dashed lines for X-ray and optical emissions in the left and middle panels of figure 6.8, respectively.

This result accounts for very long central engine activity. For an analogous analysis in the framework of the BH accretion concerning the explanation of the LT correlation, see section 6.1.1. Given that later deceleration time suggests slower ejecta and a dimmer emission, Liang et al (2010) concluded that the external shock model recovers the $L_{O,\text{peak}} - T_{O,\text{peak}}$ anti-correlation well.

Panaitescu and Vestrand (2008) claimed that the peaky afterglow phases (with $L_{O,a} \propto T_{O,a}^{-1}$) are phenomena observed slightly outside the cone of view, while the plateau afterglow phases are off-axis phenomena caused by the angular structure of the jet. Later, Panaitescu and Vestrand (2011) proposed a theoretical framework such that the peaky and plateau afterglow phases rely on how long the energy injection from the central engine lasts. In particular, the peaky afterglow phases originate from the impulsive ejecta with a small range of Γ, while the plateau afterglow phases stem from a distribution of initial Γ which feeds the energy injection up to 10^5 s.

Finally, Li et al (2012) claimed that optical flares and the optical shallow decay segments are both described by late GRB central engine activities, which can be irregular (in the case of flares) or constant (in the case of internal plateaus). Usually, after the external plateau phases a normal decay with $\alpha_{X,a}$ around -1 is encountered. These plateau phases are maybe created from an external shock with the shallow decay part produced by continuous energy injection pumping into the blast wave (Rees and Mészáros 1998, Dai and Lu 1998, Sari and Mészáros 2000, Zhang and Mészáros 2001). On the other hand, the internal plateau phases, discovered for the first time by Troja et al (2007) in GRB 070110 and later investigated statistically by Liang et al (2007), are characterized by a much steeper decay ($\alpha_{X,a}$ steeper than -3)

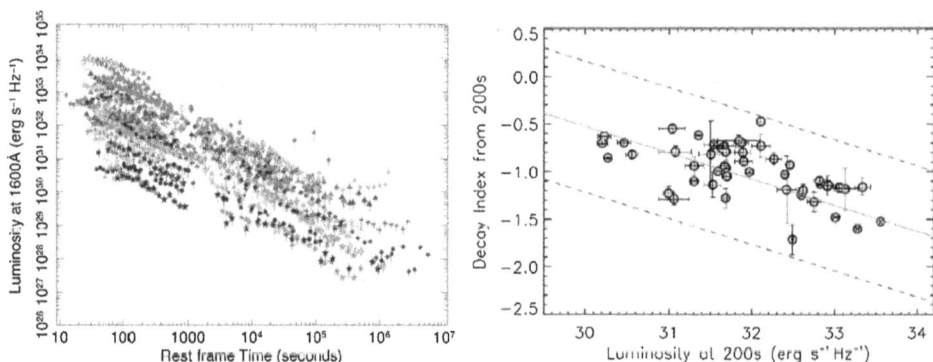

Figure 6.10. Left panel: the light curves in the optical range of the GRB dataset (Reproduced from Oates *et al* (2012).) Right panel: the $\log L_{O,\,200s}-\alpha_{O,\,>200s}$ correlation with the best-fit line displayed by a red solid line and the 3σ region indicated by a blue dashed line. (Reproduced from Oates *et al* (2012). By permission of Oxford University Press on behalf of the Royal Astronomical Society.)

after the plateau phase. These internal plateau phases are highly likely fuelled by internal dissipation of a late emission. In conclusion, the afterglow phase is well modelled by a mixture of internal and external components.

6.2 The $L_{O,200s}-\alpha_{O,>200s}$ correlation and its physical interpretation

A correlation between $\log L_{O,200s}$ and $\alpha_{O,>200s}$ (for definitions see section 3.2) was pointed out for the first time by Oates *et al* (2012) (see the right panel of figure 6.10), with a dataset of 48 UVOT LGRB light curves at 1600 Å (see the left panel of figure 6.10). This correlation is represented in the following way:

$$\log L_{O,200s} = (28.08 \pm 0.13) - (3.636 \pm 0.004) \times \alpha_{O,>200s}, \qquad (6.7)$$

with $\log L_{O,200s}$ in erg s^{-1} Hz^{-1}, $\rho = -0.58$ and a significance of 99.998% (4.2σ). Therefore, the more luminous the GRB, the faster the afterglow decay. To recover the light curves used for the analysis, they employed the following selection criteria to ensure a high S/N of the UVOT light curve: the optical/UV light curve is detected in the V filter with a magnitude $\leqslant 17.8$, UVOT observation begins during the first 400 s from the BAT trigger and the afterglow phase should last at least 10^5 s after the trigger (Oates *et al* 2009). Their outcomes showed that this correlation significantly depends on the discrepancies in the observing angle and in the rate of the energy emission from the central engine.

Additionally, Oates *et al* (2015) compared the same correlation in the optical and in X-ray bands, using the same dataset as Oates *et al* (2012). Their main result was the achievement of a similar value for the slope *b* of both correlations in the optical range. As a further step, studying the connection between $\alpha_{X,>200s}$ and $\alpha_{O,>200s}$ (for definitions see section 3.2), Oates *et al* (2015) claimed that the best-fit correlation is $\alpha_{X,>200s} = \alpha_{O,>200s} - 0.25$ (see the left panel of figure 6.11). Even if some correspondences between optical and X-ray components of GRBs were pointed out from this work, the outcomes were not in agreement with that by Urata *et al* (2007). Indeed,

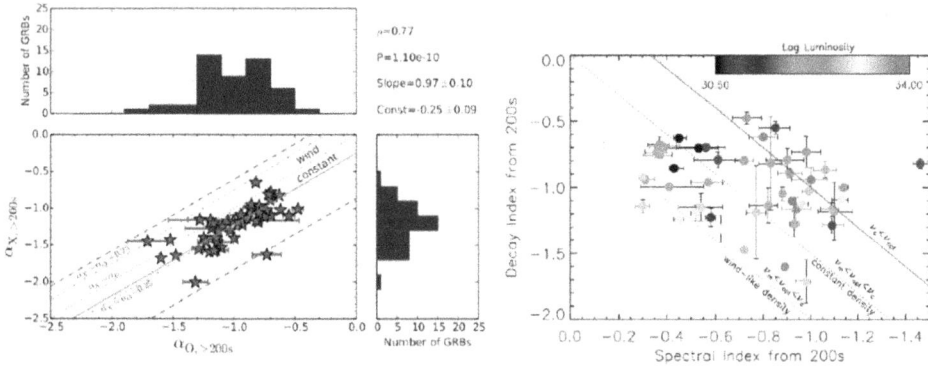

Figure 6.11. Left panel: the $\alpha_{O,\,>200s} - \alpha_{X,\,>200s}$ correlation with the best-fit line represented by a red solid line and the 3σ region indicated by blue dashed lines. The closure relations are also displayed. The $\alpha_{O,>200s} = \alpha_{X,>200s}$ correlation is given by the pink dotted line, while the $\alpha_{X,>200s} = \alpha_{O,>200s} \pm 0.25$ correlation is displayed by the light blue dotted-dashed lines. ρ, P and the best-fit coefficients are shown in the top right-hand corner (Reproduced from Oates *et al* (2015). By permission of Oxford University Press on behalf of the Royal Astronomical Society.) Right panel: the $\alpha_{O,\,>200s} - \beta_{O,\,>200s}$ correlation for 48 GRBs with different log $L_{O,200s}$ displayed by a colour scale and three closure relations indicated by lines. (Reproduced from Oates *et al* (2012). By permission of Oxford University Press on behalf of the Royal Astronomical Society.)

Urata *et al* (2007) found many outliers in the study of the correlation between $\alpha_{X,>200s}$ and $\alpha_{O,>200s}$ in the context of the external shock model. Employing 237 *Swift* LGRBs, Racusin *et al* (2016) focused on a comparable correlation in X-rays. They retrieved a $b = -0.27 \pm 0.04$ and a convincing link between the optical and X-ray components of GRBs. The Monte Carlo simulations and the statistical tests recovered the log $L_{O,\,200s} - \alpha_{O,\,>200s}$ correlation by Oates *et al* (2012). Finally, a possible association with its equivalent in X-rays, the LT correlation, was displayed, suggesting a common emission process describing them both. In table 6.3 a summary of the correlations presented in this section is given.

To explain the log $L_{O,\,200s} - \alpha_{O,\,>200s}$ correlation, Oates *et al* (2012) investigated different possibilities. In the first, considering the collision of the jet with the external medium, $\alpha_{O,>200s}$ is not a fixed quantity and all the optical afterglow phases are generated from a single closure relation with $\alpha_{O,>200s}$ and $\beta_{O,>200s}$ (for definitions see section 3.2) linearly correlated. As a consequence, a correlation between log $L_{O,200s}$ and $\beta_{O,>200s}$ should be found. However, from the observations $\alpha_{O,>200s}$ and $\beta_{O,>200s}$ are weakly related (see the right panel of figure 6.11) and no correlation between $\beta_{O,>200s}$ and log $L_{O,200s}$ is recovered. For these reasons, this theoretical interpretation is ruled out as a possible explanation of the log $L_{O,\,200s} - \alpha_{O,\,>200s}$ correlation.

For an alternative description, they considered that the log $L_{O,200s} - \alpha_{O,>200s}$ correlation originates from a few closure relations (see the lines in the right panel of figure 6.11). But, in this case the $\alpha_{O,>200s}$ and $\beta_{O,>200s}$ values with similar luminosities did not cluster around a specific closure relation. Therefore, the standard model also does not describe the log $L_{O,\,200s} - \alpha_{O,\,>200s}$ correlation well. They concluded that the afterglow model should be more complicated than the one

Table 6.3. Summary of the correlations presented in this section. The correlation in log scale is listed in the first column, the authors are presented in the second and the size of the dataset is displayed in the third. The correlation slope and normalization are indicated in the fourth and fifth columns, while the correlation coefficient and the chance probability, P, are shown in the last two columns. (Table from Dainotti and Del Vecchio (2017).)

Correlations	Author	N	Slope	Norm	Corr. coeff.	P
$L_{O,\,200s} - \alpha_{O,\,>200s}$	Oates *et al* (2012)	48	-3.636 ± 0.004	28.08 ± 0.13	-0.58	2×10^{-4}
	Oates *et al* (2015)	48	-3.636 ± 0.004	28.08 ± 0.13	-0.58	2×10^{-4}
$L_{X,\,200s} - \alpha_{X,\,>200s}$	Racusin *et al* (2016)	237	-0.27 ± 0.04	-6.99 ± 1.11	0.59	10^{-6}

assumed in previous works. It is plausible that still unknown physical features rule the radiation process and the decay rate.

For this reason, Oates *et al* (2012) indicated two further options: the first claimed that the log $L_{O,\,200s} - \alpha_{O,\,>200s}$ correlation is associated with some characteristics of the central engine which control the rate of energy emission. Thus, for weaker afterglow phases, the energy is radiated more slowly. The second option stated that the log $L_{O,\,200s} - \alpha_{O,\,>200s}$ correlation may be caused by different observing angles. In this scenario faster and more luminous decaying light curves are seen by the observers at smaller viewing angles.

In conclusion, Dainotti *et al* (2013a) claimed that the log $L_{O,\,200s} - \alpha_{O,\,>200s}$ and the LT correlations are connected and that it would be important to point out how they are associated, as well as the physical mechanisms leading to both of them. With this aim, Oates *et al* (2015) investigated the observed correlations and the simulated ones, finding that the log $L_{O,\,200s} - \alpha_{O,\,>200s}$ correlation and the equivalent in the X-ray band are compatible. This result indicated a similar physical process for both correlations.

References

Bernardini M G, Margutti R, Zaninoni E and Chincarini G 2012 A universal scaling for short and long gamma-ray bursts: $E_{X,\text{iso}} - E_{3,\text{iso}} - E_{\text{pk}}$ *Mon. Not. R. Astron. Soc.* **425** 1199–204

Bloom J S, Frail D A and Sari R 2001 The prompt energy release of gamma-ray bursts using a cosmological k-correction *Astronom. J.* **121** 2879–88

Boër M and Gendre B 2000 Evidences for two gamma-ray burst afterglow emission regimes *Astron. Astrophys.* **361** L21–4

Cannizzo J K and Gehrels N 2009 A new paradigm for gamma-ray bursts: long-term accretion rate modulation by an external accretion disk *Astrophys. J.* **700** 1047–58

Cannizzo J K, Troja E and Gehrels N 2011 Fall-back disks in long and short gamma-ray bursts *Astrophys. J.* **734** 35

Cardone V F, Capozziello S and Dainotti M G 2009 An updated gamma-ray bursts Hubble diagram *Mon. Not. R. Astron. Soc.* **400** 775–90

Cardone V F, Dainotti M G, Capozziello S and Willingale R 2010 Constraining cosmological parameters by gamma-ray burst x-ray afterglow light curves *Mon. Not. R. Astron. Soc.* **408** 1181–6

D'Agostini G 2005 Fits, and especially linear fits, with errors on both axes, extra variance of the data points and other complications, arXiv:physics/0511182

Dai Z G and Lu T 1998 Gamma-ray burst afterglows and evolution of postburst fireballs with energy injection from strongly magnetic millisecond pulsars *Astron. Astrophys.* **333** L87–90

Dainotti M, Petrosian V, Willingale R, O'Brien P, Ostrowski M and Nagataki S 2015 Luminosity–time and luminosity–luminosity correlations for GRB prompt and afterglow plateau emissions *Mon. Not. R. Astron. Soc.* **451** 3898–908

Dainotti M G and Del Vecchio R 2017 Gamma ray burst afterglow and prompt–afterglow relations: an overview *New Astron. Rev.* **77** 23–61

Dainotti M G, Cardone V F and Capozziello S 2008 A time–luminosity correlation for γ-ray bursts in the x-rays *Mon. Not. R. Astron. Soc.* **391** L79–83

Dainotti M G, Willingale R, Capozziello S, Fabrizio Cardone V and Ostrowski M 2010 Discovery of a tight correlation for gamma-ray burst afterglows with 'canonical' light curves *Astrophys. J.* **722** L215–9

Dainotti M G, Fabrizio Cardone V, Capozziello S, Ostrowski M and Willingale R 2011 Study of possible systematics in the L^{*}_{X}–T^{*}_{a} correlation of gamma-ray bursts *Astrophys. J.* **730** 135

Dainotti M G, Cardone V F, Piedipalumbo E and Capozziello S 2013a Slope evolution of GRB correlations and cosmology *Mon. Not. R. Astron. Soc.* **436** 82–8

Dainotti M G, Petrosian V, Singal J and Ostrowski M 2013b Determination of the intrinsic luminosity time correlation in the x-ray afterglows of gamma-ray bursts *Astrophys. J.* **774** 157

Dainotti M *et al* 2016 A fundamental plane for long gamma-ray bursts with x-ray plateaus *Astrophys J. Lett.* **825** L20

Dainotti M *et al* 2017 A study of the gamma-ray burst fundamental plane *Astrophys. J.* **848** 88

Dainotti M G, Nagataki S, Maeda K, Postnikov S and Pian E 2017 A study of gamma ray bursts with afterglow plateau phases associated with supernovae *Astron. Astrophys.* **600** A98

Dall'Osso S, Stratta G, Guetta D, Covino S, De Cesare G and Stella L 2011 Gamma-ray bursts afterglows with energy injection from a spinning down neutron star *Astron. Astrophys.* **526** A121

Del Vecchio R, Dainotti M G and Ostrowski M 2016 Study of GRB light-curve decay indices in the afterglow phase *Astrophys. J.* **828** 36

Efron B and Petrosian V 1992 A simple test of independence for truncated data with applications to redshift surveys *Astrophys. J.* **399** 345–52

Ghisellini G, Nardini M, Ghirlanda G and Celotti A 2009 A unifying view of gamma-ray burst afterglows *Mon. Not. R. Astron. Soc.* **393** 253–71

Gompertz B P, O'Brien P T, Wynn G A and Rowlinson A 2013 Can magnetar spin-down power extended emission in some short GRBs? *Mon. Not. R. Astron. Soc.* **431** 1745–51

Hascoët R, Daigne F and Mochkovitch R 2014 The prompt–early afterglow connection in gamma-ray bursts: implications for the early afterglow physics *Mon. Not. R. Astron. Soc.* **442** 20–7

Ito H, Nagataki S, Matsumoto J, Lee S-H, Tolstov A, Mao J, Dainotti M and Mizuta A 2014 Spectral and polarization properties of photospheric emission from stratified jets *Astrophys. J.* **789** 159

Kumar P, Narayan R and Johnson J L 2008 Properties of gamma-ray burst progenitor stars *Science* **321** 376

Leventis K, Wijers R A M J and van der Horst A J 2014 The plateau phase of gamma-ray burst afterglows in the thick-shell scenario *Mon. Not. R. Astron. Soc.* **437** 2448–60

Li L *et al* 2012 A comprehensive study of gamma-ray burst optical emission. I. Flares and early shallow-decay component *Astrophys. J.* **758** 27

Liang E and Zhang B 2006a Identification of two categories of optically bright gamma-ray bursts *Astrophys. J.* **638** L67–70

Liang E-W, Zhang B-B and Zhang B 2007 A comprehensive analysis of *Swift* XRT data. II. Diverse physical origins of the shallow decay segment *Astrophys. J.* **670** 565–83

Liang E-W, Yi S-X, Zhang J, Lü H-J, Zhang B-B and Zhang B 2010 Constraining gamma-ray burst initial Lorentz factor with the afterglow onset feature and discovery of a tight Γ_0–$E_{\mathrm{gamma,iso}}$ correlation *Astrophys. J.* **725** 2209–24

Lindner C C, Milosavljević M, Couch S M and Kumar P 2010 Collapsar accretion and the gamma-ray burst x-ray light curve *Astrophys. J.* **713** 800–15

Lü H-J and Zhang B 2014 A test of the millisecond magnetar central engine model of gamma-ray bursts with *Swift* data *Astrophys. J.* **785** 74

Lyons N, O'Brien P T, Zhang B, Willingale R, Troja E and Starling R L C 2010 Can x-ray emission powered by a spinning-down magnetar explain some gamma-ray burst light-curve features? *Mon. Not. R. Astron. Soc.* **402** 705–12

Mangano V, Sbarufatti B and Stratta G 2012 Extending the plateau luminosity–duration anticorrelation *Mem. Soc. Astron. Italiana Suppl.* **21** 143

Melandri A *et al* 2008 The early-time optical properties of gamma-ray burst afterglows *Astrophys. J.* **686** 1209–30

Melandri A *et al* 2014 Optical and x-ray rest-frame light curves of the BAT6 sample *Astron. Astrophys.* **565** A72

Nardini M, Ghisellini G, Ghirlanda G, Tavecchio F, Firmani C and Lazzati D 2006 Clustering of the optical-afterglow luminosities of long gamma-ray bursts *Astron. Astrophys.* **451** 821–33

Nardini M, Ghisellini G and Ghirlanda G 2008a Optical afterglow luminosities in the *Swift* epoch: confirming clustering and bimodality *Mon. Not. R. Astron. Soc.* **386** L87–91

Oates S R *et al* 2009 A statistical study of gamma-ray burst afterglows measured by the *Swift* Ultraviolet Optical Telescope *Mon. Not. R. Astron. Soc.* **395** 490–503

Oates S R, Page M J, De Pasquale M, Schady P, Breeveld A A, Holland S T, Kuin N P M and Marshall F E 2012 A correlation between the intrinsic brightness and average decay rate of *Swift*/UVOT gamma-ray burst optical/ultraviolet light curves *Mon. Not. R. Astron. Soc.* **426** L86–90

Oates S R, Racusin J L, De Pasquale M, Page M J, Castro-Tirado A J, Gorosabel J, Smith P J, Breeveld A A and Kuin N P M 2015 Exploring the canonical behaviour of long gamma-ray bursts using an intrinsic multiwavelength afterglow correlation *Mon. Not. R. Astron. Soc.* **453** 4121–35

Panaitescu A and Vestrand W T 2008 Taxonomy of gamma-ray burst optical light curves: identification of a salient class of early afterglows *Mon. Not. R. Astron. Soc.* **387** 497–504

Panaitescu A and Vestrand W T 2011 Optical afterglows of gamma-ray bursts: peaks, plateaus and possibilities *Mon. Not. R. Astron. Soc.* **414** 3537–46

Postnikov S, Dainotti M G, Hernandez X and Capozziello S 2014 Nonparametric study of the evolution of the cosmological equation of state with SNe Ia, BAO, and high-redshift GRBs *Astrophys. J.* **783** 126

Racusin J L, Oates S R, de Pasquale M and Kocevski D 2016 A correlation between the intrinsic brightness and average decay rate of gamma-ray burst x-ray afterglow light curves *Astrophys. J.* **826** 45

Rea N, Gullón M, Pons J A, Perna R, Dainotti M G, Miralles J A and Torres D F 2015 Constraining the GRB-magnetar model by means of the galactic pulsar population *Astrophys. J.* **813** 92

Rees M J and Mészáros P 1998 Refreshed shocks and afterglow longevity in gamma-ray bursts *Astrophys. J.* **496** L1–4

Rowlinson A and O'Brien P 2012 Energy injection in short GRBs and the role of magnetars *Gamma-Ray Bursts 2012 Conference (GRB 2012)*

Rowlinson A, O'Brien P T, Metzger B D, Tanvir N R and Levan A J 2013 Signatures of magnetar central engines in short GRB light curves *Mon. Not. R. Astron. Soc.* **430** 1061–87

Rowlinson A, Gompertz B P, Dainotti M, O'Brien P T, Wijers R A M J and van der Horst A J 2014 Constraining properties of GRB magnetar central engines using the observed plateau luminosity and duration correlation *Mon. Not. R. Astron. Soc.* **443** 1779–87

Sari R and Mészáros P 2000 Impulsive and varying injection in gamma-ray burst afterglows *Astrophys. J.* **535** L33–7

Stratta G, Dainotti M G, Dall'Osso S, Hernandez X and De Cesare G 2018 On the magnetar origin of the GRBs presenting x-ray afterglow plateaus *Astrophys. J.* **869** 155

Sultana J, Kazanas D and Fukumura K 2012 Luminosity correlations for gamma-ray bursts and implications for their prompt and afterglow emission mechanisms *Astrophys. J.* **758** 32

Sultana J, Kazanas D and Mastichiadis A 2013 The supercritical pile gamma-ray burst model: the GRB afterglow steep decline and plateau phase *Astrophys. J.* **779** 16

Troja E *et al* 2007 *Swift* observations of GRB 070110: an extraordinary x-ray afterglow powered by the central engine *Astrophys. J.* **665** 599–607

Urata Y *et al* 2007 Testing the external-shock model of gamma-ray bursts using the late-time simultaneous optical and x-ray afterglows *Astrophys. J.* **668** L95–8

van Eerten H 2014a Self-similar relativistic blast waves with energy injection *Mon. Not. R. Astron. Soc.* **442** 3495–510

van Eerten H J 2014b Gamma-ray burst afterglow plateau break time–luminosity correlations favour thick shell models over thin shell models *Mon. Not. R. Astron. Soc.* **445** 2414–23

Willingale R *et al* 2007 Testing the standard fireball model of gamma-ray bursts using late x-ray afterglows measured by *Swift Astrophys. J.* **662** 1093–1110

Yamazaki R 2009 Prior emission model for x-ray plateau phase of gamma-ray burst afterglows *Astrophys. J.* **690** L118–21

Yi S X, Dai Z G, Wu X F and Wang F Y 2014 x-ray afterglow plateaus of long gamma-ray bursts: further evidence for millisecond magnetars, arXiv: 1401.1601

Zaninoni E, Bernardini M G, Margutti R, Oates S and Chincarini G 2013 Gamma-ray burst optical light-curve zoo: comparison with x-ray observations *Astron. Astrophys.* **557** A12

Zhang B and Mészáros P 2001 Gamma-ray burst afterglow with continuous energy injection: signature of a highly magnetized millisecond pulsar *Astrophys. J.* **552** L35–8

Chapter 7

Correlations between prompt and afterglow parameters

As described in previous sections, the plateau phase is still a mystery and the correlations connected to it still need to be studied and fully understood. Thus, many models have been developed. To better understand this plateau phase the investigation of the relationship between plateau and prompt phases is needed. Here, the correlations between prompt–afterglow physical parameters are presented, providing a more exhaustive description of the plateau phase.

7.1 The $E_{X,\mathrm{afterglow}}$–$E_{\gamma,\mathrm{prompt}}$ correlation and its physical interpretation

Employing 107 *Swift* GRBs, W07 investigated the correlation between $F_{\gamma,\mathrm{prompt}}$ and $F_{X,\mathrm{afterglow}}$ (see the upper left panel of figure 7.1), measured in the BAT and XRT bands and in the XRT band, respectively. For definitions of these quantities see section 3.2. In the case of measured redshift, the isotropic energies ($E_{\gamma,\mathrm{prompt}}$ and $E_{X,\mathrm{afterglow}}$) were studied supposing a cosmological model given by $H_0 = 71$ km s^{-1} Mpc^{-1}, $\Omega_\Lambda = 0.73$ and $\Omega_M = 0.27$ (see the upper right panel of figure 7.1).

Contemporaneously, Liang *et al* (2007), with 53 LGRBs, analysed the $E_{\gamma,\mathrm{prompt}}$–$E_{X,\mathrm{afterglow}}$ correlation obtaining $b = 1 \pm 0.16$ (see the bottom left panel of figure 7.1). Later, choosing distinct energy ranges from those employed in Liang *et al* (2007, 2010) and Panaitescu and Vestrand (2011), they examined the same correlation with datasets composed of 32 and 37 GRBs, respectively. From their analysis they found slopes of $b = 0.76 \pm 0.14$ and $b = 1.18$, respectively. These correlations are represented in the left and middle panels of figure 7.2.

From the analysis of 43 SGRBs and 232 *Swift* GRBs with measured spectroscopic redshifts, respectively, Rowlinson *et al* (2013) and Grupe *et al* (2013) claimed an $E_{X,\mathrm{afterglow}}$–$E_{\gamma,\mathrm{prompt}}$ correlation with $b \sim 1$ in agreement with previous results (see the left and middle panels of figure 7.3). To put some limits on the ratio of $E_{X,\mathrm{afterglow}}$

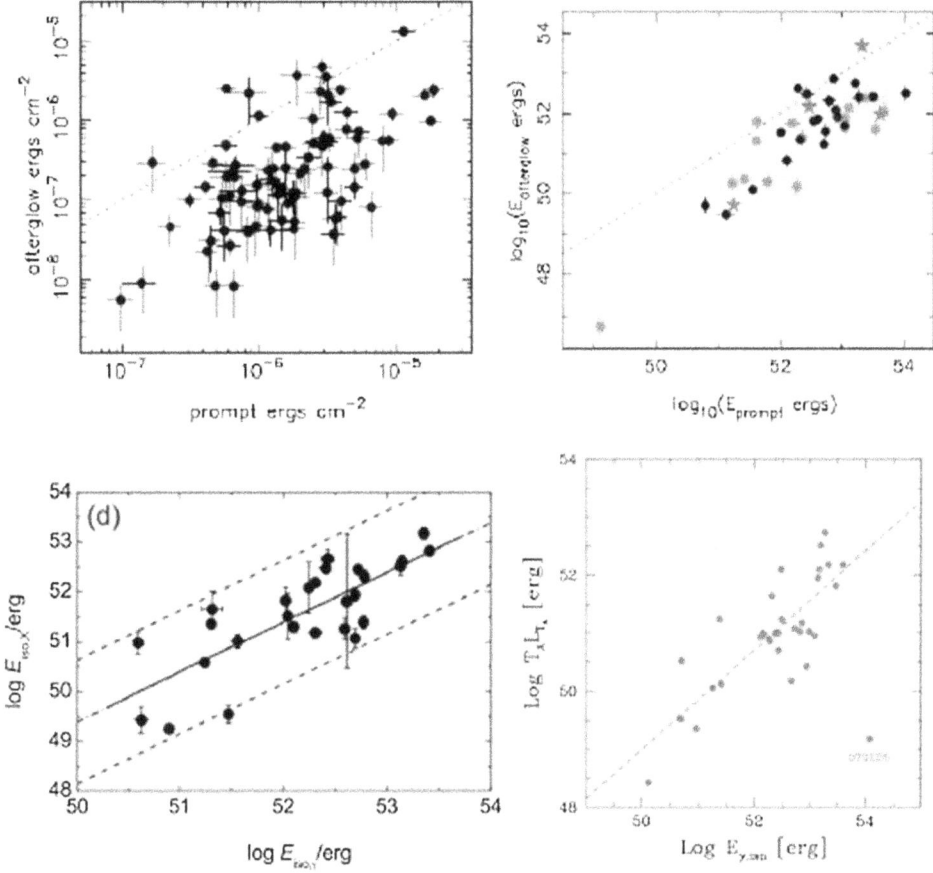

Figure 7.1. Upper left panel: the correlation between prompt and afterglow fluences obtained from the BAT flux and the XRT flux, respectively. The points where the fluences have the same values are shown by a dotted line. (Reproduced with permission from Willingale *et al* (2007). Copyright 2007 The American Astronomical Society.) Upper right panel: the $\log E_{\gamma,\,\mathrm{prompt}} - \log E_{X,\,\mathrm{afterglow}}$ correlation. GRBs in the pre-jet break area of the $\beta_{X,a} - \alpha_{X,a}$ plane are represented by dots, GRBs in the post-jet break area of the $\beta_{X,a} - \alpha_{X,a}$ plane are indicated by stars, while GRBs below the pre-jet break area of the $\beta_{X,a} - \alpha_{X,a}$ plane are displayed by squares. $\log E_{\gamma,\mathrm{prompt}} = \log E_{X,\mathrm{afterglow}}$ is shown by a dotted line. (Reproduced with permission from Willingale *et al* (2007). Copyright 2007 The American Astronomical Society.) Bottom left panel: the $\log E_{\gamma,\,\mathrm{prompt}} - \log E_{X,\,\mathrm{afterglow}}$ ($E_{\gamma,\mathrm{iso}}$ and $E_{X,\mathrm{iso}}$, respectively, in the plot) correlation with the best-fit line represented by a solid line and the 2σ region displayed by a dashed line. (Reproduced with permission from Liang *et al* (2007). Copyright 2007 The American Astronomical Society.) Bottom right panel: the $\log E_{\gamma,\,\mathrm{prompt}} - \log E_{X,\,\mathrm{plateau}}$ correlation ($E_{\gamma,\mathrm{iso}}$ and $T_a L_{Ta}$, respectively, in the figure) with the best-fit indicated by a dashed line. (Reproduced from Ghisellini *et al* (2009). By permission of Oxford University Press on behalf of the Royal Astronomical Society.)

to $E_{\gamma,\mathrm{prompt}}$, Dainotti *et al* (2015) also studied this correlation with a dataset of 123 LGRBs (see the right panel of figure 7.3). On the other hand, using the X-ray plateau energy, $E_{X,\mathrm{plateau}}$, as an evaluation of $E_{X,\mathrm{afterglow}}$, Ghisellini *et al* (2009) examined an analogous correlation for a set of 33 LGRBs. They obtained $b = 0.86$ (see the bottom right panel of figure 7.1).

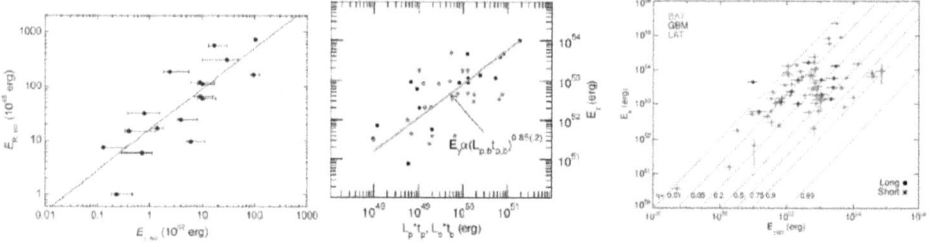

Figure 7.2. Left panel: the $E_{\gamma,\mathrm{prompt}}$–$E_{O,\mathrm{afterglow}}$ correlation ($E_{\gamma,\mathrm{iso}}$ and $E_{R,\mathrm{iso}}$, respectively, in the plot) for the optically selected dataset with the best-fit indicated by a line. (Reproduced with permission from Liang *et al* (2010). Copyright 2010 The American Astronomical Society.) Middle panel: the $E_{\gamma,\mathrm{prompt}}$–$E_{O,\mathrm{afterglow}}$ correlation ($E_{\gamma,\mathrm{iso}}$ and $L_p \times t_p$, respectively, in the plot). Afterglow phases with peaks in the optical band are indicated by black symbols, optical plateau phases are represented by red symbols and afterglows of uncertain type are displayed by open circles. For the whole sample of 37 afterglow phases $r(\log E_{X,\mathrm{afterglow}}, \log E_{\gamma,\mathrm{prompt}}) = 0.66$ and $P = 10^{-5.3}$ were obtained (Reproduced from Panaitescu and Vestrand (2011. By permission of Oxford University Press on behalf of the Royal Astronomical Society.) Right panel: the $E_{k,\mathrm{aft}}$–$E_{\gamma,\mathrm{prompt}}$ ($E_{\gamma,\mathrm{iso}}$ in the plot) correlation with distinct values of the efficiency shown by dashed lines. $E_{\gamma,\mathrm{prompt}}$ for LAT GRBs appears higher than that for BAT and GBM GRBs, while $E_{k,\mathrm{aft}}$ has an ordinary value. Thus, the efficiency for LAT GRBs will be greater than that for BAT and GBM GRBs. (Reproduced with permission from Racusin *et al* (2011). The American Astronomical Society.)

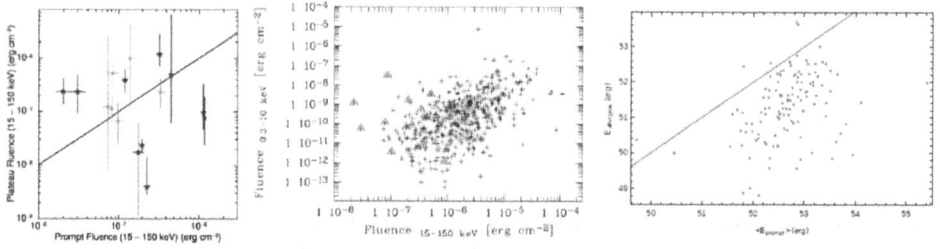

Figure 7.3. Left panel: the prompt fluence versus the X-ray fluence correlation where GRBs possessing two or more breaks in their light curves are indicated by blue stars, and GRBs with one break are represented by green circles. The equality line between the values of the shallow decay phase fluence and the prompt fluence is displayed by a black line. (Reproduced from Rowlinson *et al* (2013). By permission of Oxford University Press on behalf of the Royal Astronomical Society.) Middle panel: XRT fluence–BAT fluence correlations with triangles indicating SGRBs and circles representing high-redshift ($z > 3.5$) GRBs. (Reproduced with permission from Grupe *et al* (2013). Copyright 2013 the American Astronomical Society.) Right panel: the $\langle \log E_{\gamma,\mathrm{prompt}} \rangle$–$\log E_{X,\mathrm{afterglow}}$ correlation for a sample of 123 LGRBs with $\log E_{\gamma,\mathrm{prompt}} = \log E_{X,\mathrm{afterglow}}$ indicated by a solid line. (Reproduced from Dainotti *et al* (2015a). By permission of Oxford University Press on behalf of the Royal Astronomical Society.)

Furthermore, Ghisellini *et al* (2009) analysed the correlation between $E_{\gamma,\mathrm{prompt}}$ and $E_{k,\mathrm{aft}}$ with the same dataset, for definitions see section 3.2. They found $b = 0.42$. Analogously, this correlation was also examined by Racusin *et al* (2011), with a sample of 69 GRBs. In their analysis they selected different values for the efficiency to constrain $E_{k,\mathrm{aft}}$ and $E_{\gamma,\mathrm{prompt}}$ (see the right panel of figure 7.2). The main purpose of the use of this correlation was to test the difference in the detection limits of many instruments and to study the transfer of kinetic energy in the prompt radiation. Among the works mentioned above, the correlation presented by Racusin *et al* (2011) is considered the most robust.

In conclusion, the $E_{X,\text{afterglow}}-E_{\gamma,\text{prompt}}$ correlation, investigated by Liang *et al* (2007) and confirmed by Rowlinson *et al* (2013), Grupe *et al* (2013) and Dainotti *et al* (2015), is useful for relating the energies in the prompt and the afterglow phases. The $E_{k,\text{aft}}-E_{\gamma,\text{prompt}}$ correlation analysed by Ghisellini *et al* (2009) and confirmed by Racusin *et al* (2011) can help to examine the efficiency of the emission processes in the transition from the prompt to the afterglow phases. Additionally, this correlation is able to describe the relationship between these two phases well. Ghisellini *et al* (2009) and Racusin *et al* (2011) concluded that the transferred kinetic energy from the prompt to the afterglow phases is approximately 10% for a dataset of BAT GRBs. This outcome is compatible with those by Zhang *et al* (2007a). In particular, the internal shock model explains well the 10% for the energy efficiency when a late energy transfer occurs from the fireball to the external medium (Zhang and Kobayashi 2005).

In table 7.1, the parameters of the correlations mentioned in this section are described.

To physically describe the $E_{X,\text{afterglow}}-E_{\gamma,\text{prompt}}$ correlation it is important to consider that the outcomes by Racusin *et al* (2011) on the efficiency of a BAT dataset are in agreement with the results of Zhang *et al* (2007a), for which ~57% of BAT bursts have energy efficiency < 10%. From the sets of GRBs detected by the GBM and LAT instruments, it was found that only 25% of the GBM GRBs and none of the LAT GRBs have efficiency < 10%. Therefore, it was proposed that the efficiency of the kinetic energy transfer to prompt emission of *Fermi* GRBs is higher than that of *Swift* GRBs.

Table 7.1. Summary of the correlations presented in this section. The correlation in log scale is listed in the first column, the authors are presented in the second and the size of the dataset is displayed in the third. The correlation slope and normalization are indicated in the fourth and fifth columns, while the correlation coefficient and the chance probability, P, are shown in the last two columns. (Table from Dainotti and Del Vecchio (2017).)

Correlations	Author	N	Slope	Norm	Corr. coeff.	P
$E_{X,\text{afterglow}}-E_{\gamma,\text{prompt}}$	Liang *et al* (2007)	53	$1.00^{+0.16}_{-0.16}$	$-0.50^{+8.10}_{-8.10}$	0.79	$< 10^{-4}$
$E_{O,\text{afterglow}}-E_{\gamma,\text{prompt}}$	Liang *et al* (2010)	32	$0.76^{+0.14}_{-0.14}$	$1.30^{+0.14}_{-0.14}$	0.82	$< 10^{-4}$
	Panaitescu and Vestrand (2011)	37	1.18		0.66	$10^{-5.3}$
$E_{X,\text{plateau}}-E_{\gamma,\text{prompt}}$	Ghisellini *et al* (2009)	33	0.86			2×10^{-7}
$E_{k,\text{aft}}-E_{\gamma,\text{prompt}}$	Ghisellini *et al* (2009)	33	0.42			10^{-3}

7.2 The $L_{X,\text{afterglow}}$–$E_{\gamma,\text{prompt}}$ correlation and its physical interpretation

Examining the observed energies of 16 SGRBs in the prompt and afterglow phases, Berger (2007) pointed out that the 80% fulfils a linear correlation between $S_{\gamma,\text{prompt}}$ and $F_{X,1\text{d}}$ in the BAT and XRT range, respectively. For definitions see section 3.2. The correlation between these quantities is the following:

$$\log F_{X,1\text{d}} \sim (1.01 \pm 0.09) \times \log S_{\gamma,\text{prompt}}, \tag{7.1}$$

with $\rho = 0.86$ and $P = 5.3 \times 10^{-5}$. Using $F_{X,11}$ instead of $F_{X,1\text{d}}$ (for definitions see section 3.2) Gehrels *et al* (2008) verified this correlation found by Berger (2007) (see figure 7.4).

Subsequently, Nysewander *et al* (2009), with 37 SGRBs and 421 LGRBs observed by *Swift*, obtained a roughly linear correlation between $F_{X,11}$ or $F_{O,11}$ and $E_{\gamma,\text{prompt}}$ (see figure 7.5; for definitions see section 3.2). An analogous correlation between $E_{\gamma,\text{prompt}}$ and $F_{O,a}$ with a steeper slope $b = 1.67$ was found by Panaitescu and Vestrand (2011), employing 37 GRBs (see the left panel of figure 7.6).

Additionally, using four long events associated with SNe and 27 'regular' energetic LGRBs ($E_{\gamma,\text{prompt}} \sim 10^{52}$–$10^{54}$ erg), Kaneko *et al* (2007) pointed out a linear correlation $L_{X,10} \propto E_{\gamma,\text{prompt}}$. For definitions see section 3.2. $L_{X,10}$ is in the energy band between 2 and 10 keV, while $E_{\gamma,\text{prompt}}$ is computed between 20 and 2000 keV (see the left panel of figure 7.7). From their analysis, the possibility of a similar efficiency value in the kinetic energy transformation for the four events associated with SNe and for the 'regular' energetic LGRBs was claimed.

This correlation was investigated for the low-luminosity versus normal luminosity GRBs. Amati *et al* (2007) investigated the correlation between $L_{X,10}$, in the range between 2 and 10 keV, and $E_{\gamma,\text{prompt}}$, in the range between 1 and 10 000 keV. From

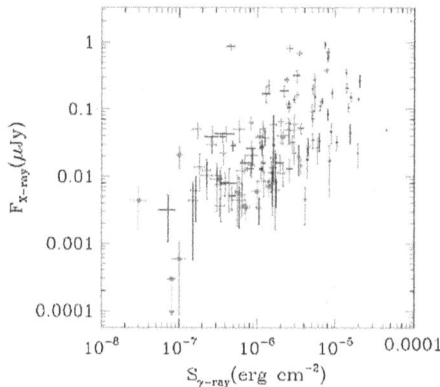

Figure 7.4. The $F_{X,11}$–$S_{\gamma,\text{prompt}}$ distribution ($F_{\text{X-ray}}$ and $S_{\gamma\text{-ray}}$, respectively, in the picture) for SGRBs (indicated by red points) and LGRBs (indicated by blue points) detected by *Swift*. The XRT $F_{X,11}$ are computed at 3 keV and the BAT $S_{\gamma,\text{prompt}}$ are detected in the whole BAT energy range. (Reproduced with permission from Gehrels *et al* (2008). Copyright 2008 The American Astronomical Society.)

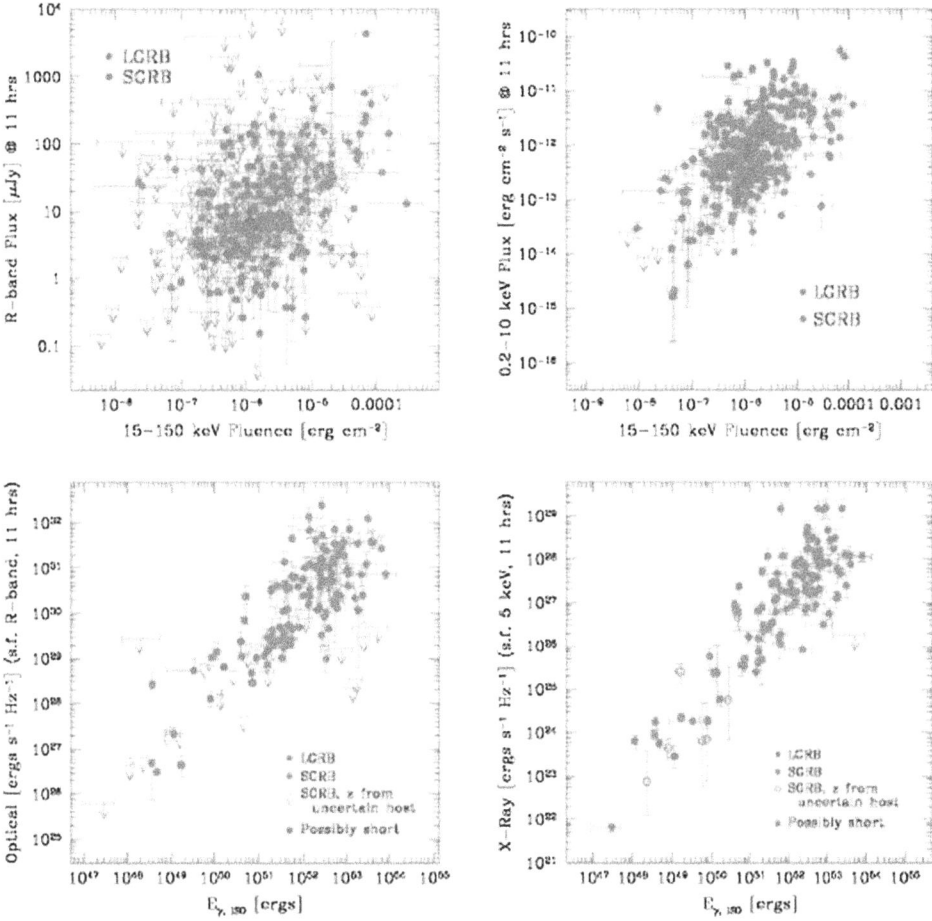

Figure 7.5. Upper left panel: the $S_{\gamma,\mathrm{prompt}}$–$F_{O,11}$ correlation using LGRBs (presented in grey) and SGRBs (presented in red). $F_{O,11}$ is corrected for galactic extinction, and $S_{\gamma,\mathrm{prompt}}$ is computed in the BAT band. Optical afterglow phases for SGRBs are not found for values of the fluence lower than 10^{-7} erg cm^{-2}, while for fluence values higher than 10^{-7} erg cm^{-2} optical afterglow phases are observed for all GRBs in the sample, except for GRB 061210. Upper right panel: the $S_{\gamma,\mathrm{prompt}}$–$F_{X,11}$ correlation using LGRBs (presented in grey) and SGRBs (presented in red). Bottom left panel: the $L_{O,11}$–$E_{\gamma,\mathrm{prompt}}$ ($E_{\gamma,\mathrm{iso}}$ in the plot) correlation with GRB 060614 and GRB 060505 indicated as 'possibly short'. $L_{O,11}$ is corrected for Galactic extinction. SGRBs with a host galaxy computed by XRT error circles are indicated by dashed upper limits. Bottom right panel: the $L_{X,11}$–$E_{\gamma,\mathrm{prompt}}$ ($E_{\gamma,\mathrm{iso}}$ in the plot) correlation with GRB 060614 and GRB 060505 indicated as 'possibly short'. SGRBs with a host galaxy computed by XRT error circles are indicated by open circles. (Reproduced with permission from Nysewander *et al* (2009). Copyright 2009 The American Astronomical Society.)

their analysis, it was concluded that this correlation is tighter ($P \sim 10^{-11}$) when sub-energetic GRBs, such as GRB 060218, GRB 980425 and GRB 031203, are taken into account (see the middle panel of figure 7.7). Sub-energetic GRBs are intrinsically dim and they can be regarded as normal cosmological GRBs. As a further step, the correlation between $L_{X,1d}$ and $E_{\gamma,\mathrm{prompt}}$ (for definitions see section 3.2) was

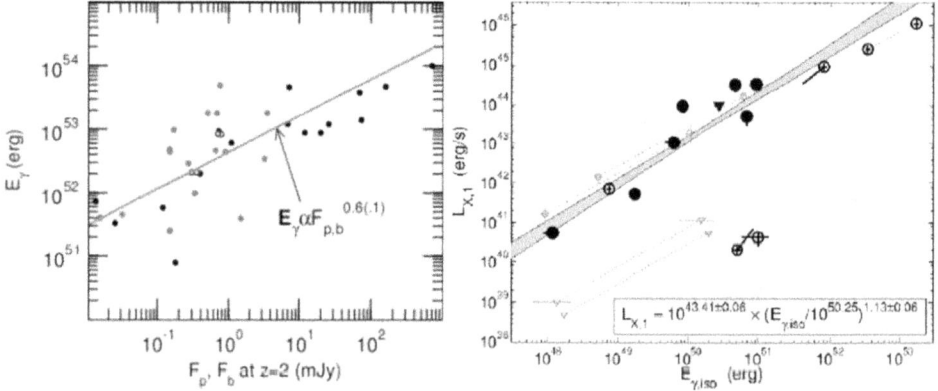

Figure 7.6. Left panel: the $E_{\gamma,\mathrm{prompt}}$–$F_{O,a}$ correlation ($E_{\gamma,\mathrm{iso}}$ and F_p, respectively, in the plot). Afterglow phases with optical peaks are indicated by black symbols, optical plateau phases are represented by red symbols and afterglow phases of unknown type are displayed by open circles. (Reproduced from Panaitescu and Vestrand (2011). By permission of Oxford University Press on behalf of the Royal Astronomical Society.) Right panel: the $L_{X,1\mathrm{d}}$–$E_{\gamma,\mathrm{prompt}}$ correlation ($E_{\gamma,\mathrm{iso}}$ in the plot). SGRBs with measured z are indicated by solid black circles, SGRBs with redshift constraints are presented by open black circles, while SGRBs without measured redshift are displayed by grey symbols connected by dotted lines. (Reproduced with permission from Berger (2007). Copyright 2007 The American Astronomical Society.)

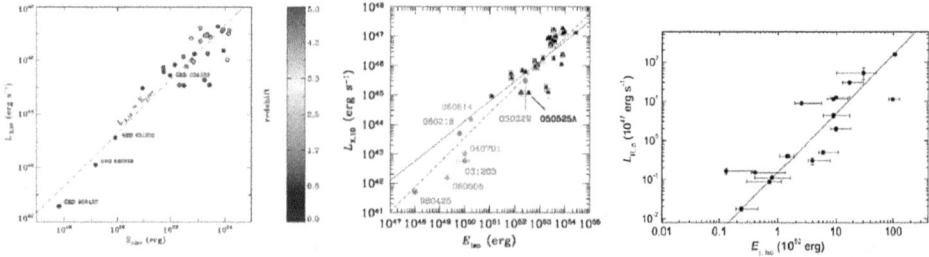

Figure 7.7. Left panel: the $L_{X,10}$–$E_{\gamma,\mathrm{prompt}}$ ($E_{\gamma,\mathrm{iso}}$ in the picture) correlation for GRBs connected with SNe. $E_{\gamma,\mathrm{prompt}}$ is computed between 20 and 2000 keV, while $L_{X,10}$ is between 2 and 10 keV. A coloured bar displays the redshift scale. (Reproduced with permission from Kaneko *et al* (2007). Copyright 2007 The American Astronomical Society.) Middle panel: the $L_{X,10}$–$E_{\gamma,\mathrm{prompt}}$ (E_{iso} in the picture) correlation. GRBs by Nousek *et al* (2006) are indicated by triangles, the three sub-energetic GRBs, GRB 980425, GRB 031203 and GRB 060218, and the GRBs connected with SNe GRB 030329 are represented by circles, while the three GRBs with measured z and deep constraints on the peak magnitude of the associated SNe, XRF 040701, GRB 060505 and GRB 060614, are displayed by diamonds. Empty triangles show GRBs with $E_{\gamma,\mathrm{prompt}}$ in the range of 1–10 000 keV calculated from $E_{\gamma,\mathrm{prompt}}$ in the 100–500 energy keV tabulated by Nousek *et al* (2006) considering an average spectral index. The best-fit calculated without sub-energetic GRBs and GRB 030329 is indicated by the dotted line, while that with sub-energetic GRBs and GRB 030329 is indicated by the dashed line. (Reproduced with permission from Amati *et al* (2007). Copyright 2007 ESO.) Right panel: the $E_{\gamma,\mathrm{prompt}}$–$L_{O,\mathrm{peak}}$ correlation ($E_{\gamma,\mathrm{iso}}$ and $L_{R,p}$, respectively, in the plot) using the optically selected sample with the best-fit indicated by the line. (Reproduced with permission after Liang *et al* (2010). Copyright 2010 The American Astronomical Society.)

examined by Berger (2007) with a sample of 13 SGRBs with known z. They recovered a slope $b = 1.13 \pm 0.16$ (see the right panel of figure 7.6).

In agreement with previous outcomes, Liang *et al* (2010) obtained with 32 *Swift* GRBs an $L_{O,\text{peak}}$–$E_{\gamma,\text{prompt}}$ correlation in the optical band with $b = 1.40 \pm 0.08$ (see the right panel of figure 7.7). Moreover, Kann *et al* (2010) obtained, using 76 LGRBs, an $L_{O,\text{1d}}$–$E_{\gamma,\text{prompt}}$ correlation with $b = 0.36$ (see the left panel of figure 7.8) confirming the results by Berger (2007).

Analogously, a correlation between $\log L_{X,a}$ and $\log E_{\gamma,\text{prompt}}$ was studied by Dainotti *et al* (2011) for a dataset of 66 *Swift* BAT+XRT LGRBs, whose light curves were taken from the Swift burst analyzer http://www.swift.ac.uk/burst_analyser/. Employing σ_E as a quantity indicating the goodness of the fit, the dataset was separated into the E4 subsample, composed of 62 LGRBs, and the E0095 subsample, formed of eight LGRBs. In the case of the E4 subsample it was found that

$$\log L_{X,a} = 28.03^{+2.98}_{-2.97} + 0.52^{+0.07}_{-0.06} \times \log E_{\gamma,\text{prompt}}, \qquad (7.2)$$

with $\rho = 0.43$ and $P = 1.4 \times 10^{-5}$, while for the E0095 subsample it was obtained that

$$\log L_{X,a} = 29.82^{+7.11}_{-7.82} + 0.49^{+0.21}_{-0.16} \times \log E_{\gamma,\text{prompt}}, \qquad (7.3)$$

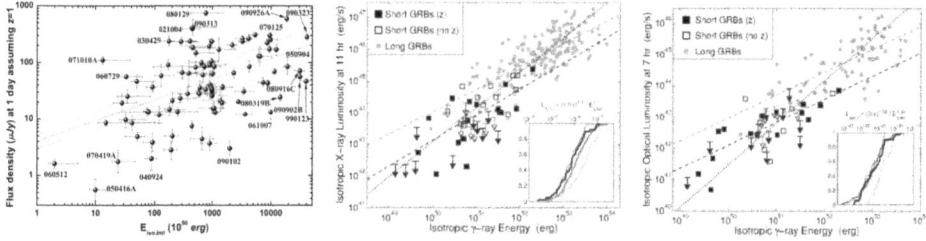

Figure 7.8. Left panel: the $F_{O,\text{1d}}$–$E_{\gamma,\text{prompt}}$ ($E_{\text{iso,bol}}$ in the picture) correlation for the optically selected sample with the best-fit line represented by a dashed line and the 3σ error indicated by a dotted line. GRB 991208, 060210, 060607A, 060906 and 080319C are not included. The first because it was only observed after few days, and the others because the detection stopped before one day. Particular GRBs are displayed with their own names. (Reproduced with permission from Kann *et al* (2010). Copyright 2010 The American Astronomical Society.) Middle panel: the $L_{X,11}$–$E_{\gamma,\text{prompt}}$ correlation using SGRBs (indicated in blue) and LGRBs (indicated in grey). SGRBs without measured redshift, but with a selected fiducial value $z = 0.75$, are represented by open markers. The best-fit for SGRBs and LGRBs are indicated by the dashed blue and red lines, respectively. The correlation predicted by the afterglow synchrotron model with $\nu_X > \nu_c$ and $p = 2.4$ ($\log L_{X,11} \propto 1.1 \times \log E_{\gamma,\text{prompt}}$) is displayed by a dotted black line. In the inset panel thick lines indicate the distribution of the quantity $\log(L_{X,11} \times (11\,\text{h})^{1.3}/E_{\gamma,\text{prompt}}^{1.1})$ for the complete dataset, while thin lines display the same distribution but for the area where values of $E_{\gamma,\text{prompt}}$ of SGRBs and LGRBs are identical. For SGRBs, $L_{X,11}$ at the same value of $E_{\gamma,\text{prompt}}$ appears clearly smaller than that for LGRBs. (Reproduced from Berger (2014).) Right panel: the $L_{O,7}$–$E_{\gamma,\text{prompt}}$ correlation using SGRBs (indicated in blue) and LGRBs (indicated in grey). The correlation predicted by the afterglow synchrotron model for $\nu_m < \nu_O < \nu_c$ and $p = 2.4$ ($\log L_{O,7} \propto 1.35 \times \log E_{\gamma,\text{prompt}}$) is represented by a dotted black line. In the inset panel the thick lines display the distribution of the quantity $\log(L_{O,7} \times (7\,\text{h})^{1.05}/E_{\gamma,\text{prompt}}^{1.35})$ for the complete dataset, while thin lines display the same distribution for the region where values of $E_{\gamma,\text{prompt}}$ for SGRBs and LGRBs are identical. For SGRBs, $L_{O,7}$ at the same value of $E_{\gamma,\text{prompt}}$ appears clearly smaller than that for LGRBs. (Reproduced with permission from Berger (2014). Copyright ESO.)

with $\rho = 0.83$ and $P = 3.2 \times 10^{-2}$. From this analysis, it was found that the subset with the smallest σ_E shows a tighter correlation. In principle this result could allow for the use of this subsample of GRBs as a standardizable candle. Furthermore, since the LT slope, b, is roughly -1, the energy reservoir of the plateau phase, $\log E_{X,\text{plateau}}$ is roughly constant. Since $\log E_{\gamma,\text{peak}}$ and $\log E_{\gamma,\text{prompt}}$ are both correlated with $\log E_{X,\text{plateau}}$, the $\log E_{\gamma,\text{peak}}-\log E_{\gamma,\text{prompt}}-\log E_{X,\text{plateau}}$ correlation is easily obtained. Bernardini *et al* (2012) investigated a variation of the $\log E_{\gamma,\text{peak}}-\log E_{\gamma,\text{prompt}}-\log E_{X,\text{plateau}}$ correlation employing $\log E_{\gamma,\text{iso}}$ of the complete light curve in the X-ray energy band. In conclusion, with datasets of 58 and 297 LGRBs from *Swift*, respectively, D'Avanzo *et al* (2012) and Margutti *et al* (2013) verified the $L_{X,a}-E_{\gamma,\text{prompt}}$ correlation, pointing out $b \sim 1$ and $\rho \approx 0.70$.

The correlations between $E_{\gamma,\text{peak}}$ and $L_{X,11}$ and $E_{\gamma,\text{peak}}$ and $L_{O,7}$ (for definitions see section 3.2) were investigated by Berger (2014) employing 70 SGRBs and 73 LGRBs. The recovered correlations were flatter than those obtained by Kann *et al* (2010) through the simulations (see the middle and right panels of figure 7.8). In addition, a tight correlation between $L_{O,200\text{s}}$ or $L_{X,200\text{s}}$ and $\log E_{\gamma,\text{prompt}}$ was found by Oates *et al* (2015) employing a dataset of 48 LGRBs (see figure 7.9 and table 7.2). From their analysis, it was concluded that this correlation allows for a deeper investigation of the spectral features of GRBs, the emission mechanism in the optical and X-ray energy ranges, and the standard afterglow model. In table 7.2, the correlations presented in this section are displayed. For the physical interpretation of the $L_{X,\text{afterglow}}-E_{\gamma,\text{prompt}}$ correlation, Gehrels *et al* (2008) stressed that $\beta_{OX,a} \approx 0.75$ identifies the optical and X-ray emissions. Nevertheless, this value of $\beta_{OX,a}$ is similar to the one for the slow cooling case, when $\gamma = 2.5$ for $\nu_m < \nu_O < \nu_X < \nu_c$. γ is the electron distribution index, ν_X indicates the X-ray frequency, ν_O represents the optical frequencies, ν_c is the cooling frequency and ν_m designates the peak frequencies of the synchrotron radiation. Oates *et al* (2015) found that the standard afterglow model anticipates the $(\log L_{O,200\text{s}}, \log L_{X,200\text{s}})-\log E_{\gamma,\text{prompt}}$ correlations. However, the slopes of the simulated and observed correlations are not compatible at $> 3\sigma$ as a consequence of the values of the efficiency, η. The scatter of the correlation will be more significant if

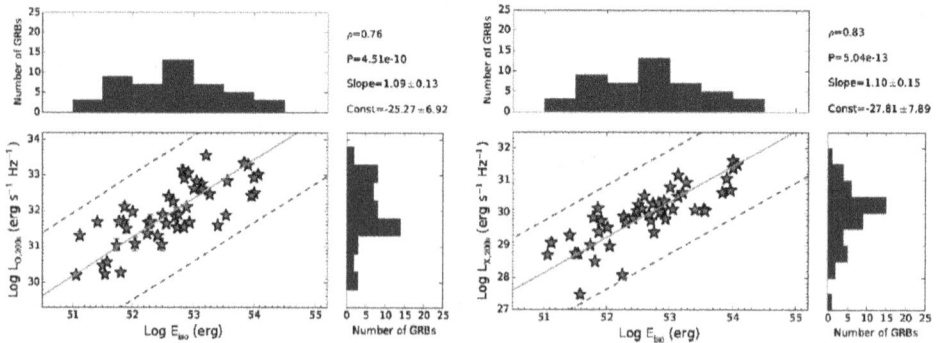

Figure 7.9. Left panel: the $\log L_{O,200\text{s}}-\log E_{\gamma,\text{prompt}}$ correlation. (Reproduced from Oates *et al* (2015).) Right panel: the $\log L_{X,200\text{s}}-\log E_{\gamma,\text{prompt}}$ correlation. (Reproduced from Oates *et al* (2015). By permission of Oxford University Press on behalf of the Royal Astronomical Society.) In both panels $E_{\gamma,\text{prompt}}$ is labelled E_{iso}.

Table 7.2. Summary of the correlations presented in this section. The correlation in log scale is listed in the first column, the authors are presented in the second and the size of the dataset is displayed in the third. The correlation slope and normalization are indicated in the fourth and fifth columns, while the correlation coefficient and the chance probability, P, are shown in the last two columns. (Table from Dainotti and Del Vecchio (2017).)

Correlations	Author	N	Slope	Norm	Corr. coeff.	P
$F_{X,1d}$–$S_{\gamma,\mathrm{prompt}}$	Berger (2007)	16	$1.01^{+0.09}_{-0.09}$		0.86	5.3×10^{-5}
$F_{X,11}$–$S_{\gamma,\mathrm{prompt}}$	Gehrels et al (2008)	111	$0.63^{+0.04}_{-0.04}$	$2.11^{+0.21}_{-0.21}$	0.53	4×10^{-9}
	Gehrels et al (2008)	10	$0.36^{+0.17}_{-0.17}$	$0.06^{+1.07}_{-1.07}$	0.35	0.31
$F_{O,11}$–$E_{\gamma,\mathrm{prompt}}$	Nysewander et al (2009)	421	~ 1			
$F_{O,11}$–$E_{\gamma,\mathrm{prompt}}$	Nysewander et al (2009)	37	~ 1			
$F_{X,11}$–$E_{\gamma,\mathrm{prompt}}$	Nysewander et al (2009)	421	~ 1			
$F_{X,11}$–$E_{\gamma,\mathrm{prompt}}$	Nysewander et al (2009)	37	~ 1			
$F_{O,a}$–$E_{\gamma,\mathrm{prompt}}$	Panaitescu and Vestrand (2011)	37	1.67		0.75	$10^{-7.3}$
$L_{X,1d}$–$E_{\gamma,\mathrm{prompt}}$	Berger (2007)	13	$1.13^{+0.16}_{-0.16}$	$43.43^{+0.20}_{-0.20}$	0.94	3.2×10^{-6}
$L_{O,\mathrm{peak}}$–$E_{\gamma,\mathrm{prompt}}$	Liang et al (2010)	32	$1.40^{+0.08}_{-0.08}$	$0.83^{+0.15}_{-0.15}$	0.87	10^{-4}
$L_{O,1d}$–$E_{\gamma,\mathrm{prompt}}$	Kann et al (2010)	76	0.36			
$L_{X,a}$–$E_{\gamma,\mathrm{prompt}}$	Dainotti et al (2011b)	62	$0.52^{+0.07}_{-0.06}$	$28.03^{+2.98}_{-2.97}$	0.43	1.4×10^{-5}
	Dainotti et al (2011b)	8	$0.49^{+0.21}_{-0.16}$	$29.82^{+7.11}_{-7.82}$	0.83	3.2×10^{-2}
	D'Avanzo et al (2012)	58	~ 1		≈ 0.70	
	Margutti et al (2013)	297	~ 1		≈ 0.70	
$L_{X,11}$–$E_{\gamma,\mathrm{prompt}}$	Berger (2014)	73	0.72	44.75		
	Berger (2014)	70	0.83	43.93		
$L_{O,7}$–$E_{\gamma,\mathrm{prompt}}$	Berger (2014)	73	0.73	43.70		
	Berger (2014)	70	0.74	42.84		
$L_{X,200s}$–$E_{\gamma,\mathrm{prompt}}$	Oates et al (2015)	48	$1.10^{+0.15}_{-0.15}$	$-27.81^{+7.89}_{-7.89}$	0.83	5.04×10^{-13}
$L_{O,200s}$–$E_{\gamma,\mathrm{prompt}}$	Oates et al (2015)	48	$1.09^{+0.13}_{-0.13}$	$-25.27^{+6.92}_{-6.92}$	0.76	4.51×10^{-10}

the efficiency range is not sufficiently small. For example, the simulations carried out assuming efficiency $\eta = 0.1$ and $\eta = 0.9$ indicated inconsistent outcomes between the simulated and observed slopes by more than 3σ.

7.3 The $L_{X,a}$–$L_{O,a}$ correlation and its physical interpretation

Jakobsson et al (2004) investigated the log $F_{O,11}$–log $F_{X,11}$ distribution in the optical R band and in the 2–10 keV energy range, respectively. Using all GRBs with an observed X-ray afterglow phase (see the left panel of figure 7.10), they recovered five

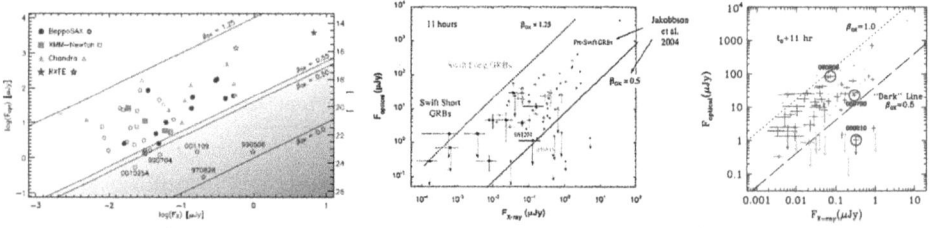

Figure 7.10. Left panel: the $\log F_{O,11} - \log F_{X,11}$ correlation (F_{opt} and F_X, respectively, in the plot) with optical observations displayed by filled symbols and upper limits indicated by open symbols. Constant $\beta_{OX,a}$ lines are displayed with their respective values. GRBs with $\beta_{OX,a} < 0.5$ are called dark bursts. (Reproduced with permission from Jakobsson *et al* (2004). Copyright 2004 The American Physical Society.) Middle panel: the $F_{X,11} - F_{O,11}$ correlation for *Swift* SGRBs and LGRBs. The pre-*Swift* GRBs and $\beta_{OX,a}$ values from Jakobsson *et al* (2004) are inserted for analogy. The grey points display LGRBs, the black points indicate SGRBs and the small black points without error bars show the pre-*Swift* GRBs. (Reproduced from Gehrels (2007). Courtesy of NASA.) Right panel: the $F_{O,11} - F_{X,11}$ correlation using a sample of *Swift* SGRBs (indicated by red dots) and LGRBs (indicated by blue dots). The three GRBs with $z > 3.9$ are presented inside a circle, while in green are displayed the pre-*Swift* GRBs from Jakobsson *et al* (2004). The line $\beta_{OX,a} = 0.5$ for dark bursts (Jakobsson *et al* 2004) and a line for $\beta_{OX,a} = 1.0$ are described. (Reproduced with permission from Gehrels *et al* (2008). Copyright 2008 The American Physical Society.)

dark bursts from 52 *BeppoSAX* bursts. Unlike earlier definitions, they denoted dark bursts as those GRBs with $\beta_{OX,a}$, smaller than 0.5, i.e. $\beta_{OX,a} \lesssim 0.5$. This definition was later employed by Nardini *et al* (2006). Their purpose was to extend the investigation of dark bursts to *Swift* data. Later, Gehrels (2007) and Gehrels *et al* (2008), with 19 SGRBs and 37 LGRBs+6 SGRBs, respectively, validated previous outcomes (see the middle and right panels of figure 7.10), recovering $b = 0.38 \pm 0.03$ for LGRBs and $b = 0.14 \pm 0.45$ for SGRBs (Gehrels *et al* 2008).

Regarding the rest-frame, Berger (2014) pointed out some resemblances between SGRBs and LGRBs, and a common average value of the ratio $\langle L_{O,7}/L_{X,11} \rangle \approx 0.08$, by investigating the correlation between $L_{O,7}$ and $L_{X,11}$ for 70 SGRBs and 73 LGRBs (see the left panel of figure 7.11). Oates *et al* (2015) enriched this analysis by examining $L_{O,200s}$ and $L_{X,200s}$ for 48 LGRBs (see the right panel of figure 7.11) and by computing $b = 0.91 \pm 0.22$. The main aim of this correlation is to investigate the GRB synchrotron spectrum and to recover some limits on the external medium for both LGRBs and SGRBs. In table 7.3 the parameters of the correlations presented in this section are shown.

For the physical interpretation of the $L_{X,a} - L_{O,a}$ correlation, Berger (2014) claimed that in the framework of the synchrotron model the correlation between $L_{O,7}$ and $L_{X,11}$ implies that often ν_c is around or greater than the X-ray frequency range. This leads to the conclusion that the external medium densities for LGRBs are usually ~ 50 times higher than those for SGRBs, thus $\nu_c \sim \nu_X$.

7.4 The $L_{X,a} - L_{\gamma,iso}$ correlation

Employing 77 *Swift* LGRBs, Dainotti *et al* (2011) investigated the relationships between the prompt emission and $\log L_{X,a}$. They obtained a correlation between

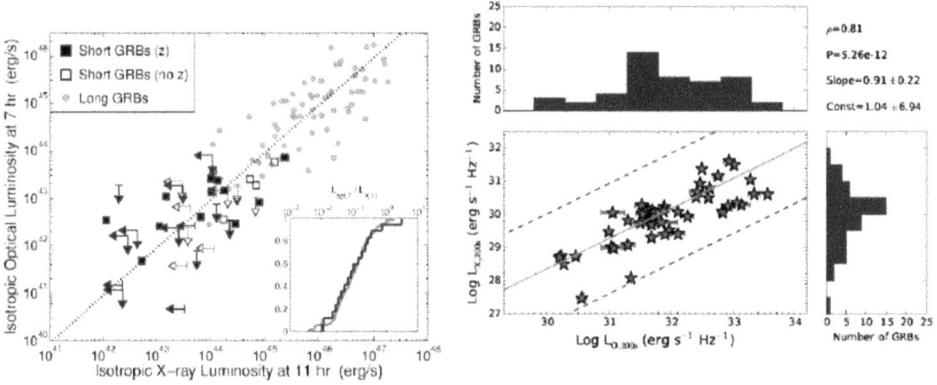

Figure 7.11. Left panel: the $L_{O,7}$–$L_{X,11}$ correlation. The correlation predicted by $\nu_X \sim \nu_c$ is represented by a dotted black line. In the inset panel the distribution of the quantity $L_{O,7}/L_{X,11}$ is displayed. These distributions are analogous for both SGRBs and LGRBs, with $L_{O,7}/L_{X,11} \lesssim 1$. This finding suggests that $\nu_X \sim \nu_c$ for SGRBs. (Reproduced from Berger (2014). Copyright ESO.) Right panel: the $\log L_{X,200s}$– $\log L_{O,200s}$ correlation with the best-fit line presented by a red solid line and the 3σ error region shown by a blue dashed line. ρ, P and the best-fit coefficients are displayed in the top right corner. (Reproduced from Oates *et al* (2015). By permission of Oxford University Press on behalf of the Royal Astronomical Society.)

Table 7.3. Summary of the correlations presented in this section. The correlation in log scale is listed in the first column, the authors are presented in the second and the size of the dataset is displayed in the third. The correlation slope and normalization are indicated in the fourth and fifth columns, while the correlation coefficient and the chance probability, P, are shown in the last two columns. (Table from Dainotti and Del Vecchio (2017).)

Correlations	Author	N	Slope	Norm	Corr. coeff.	P
$F_{X,11}$–$F_{O,11}$	Gehrels *et al* (2008)	6	0.14 ± 0.45	0.72 ± 0.94	0.06	0.68
		37	0.38 ± 0.03	1.62 ± 0.04	0.44	0.006
$L_{X,11}$–$L_{O,7}$	Berger (2014)	70	0.08			
		73	0.08			
$L_{X,200s}$– $L_{O,200s}$	Oates *et al* (2015)	48	0.91 ± 0.22	1.04 ± 6.94	0.81	5.26×10^{-12}

$\log L_{X,a}$ in the XRT range and $\log \langle L_{\gamma,\mathrm{iso}} \rangle_{45} \equiv \log(E_{\gamma,\mathrm{prompt}}/T_{45})$ in the BAT range, for definitions see section 3.2. From the sample of 77 GRBs they selected a subsample of LGRBs for which $\sigma_E < 4$ is fulfilled, similarly to what has been done for the LT correlation, see section 6.1.1. This new sample is composed of 62 LGRBs and the correlation (see the left panel of figure 7.12) associated with it is

$$\log L_{X,a} = 20.58^{+6.66}_{-6.73} + 0.67^{+0.14}_{-0.15} \times \log\langle L_{\gamma,\mathrm{iso}}\rangle_{45}, \tag{7.4}$$

with $\rho = 0.59$ and $P = 7.7 \times 10^{-8}$. Additionally, $\log L_{X,a}$ was correlated with other prompt luminosities determined by employing different timescales, such as T_{90}, T_{45} and $T_{X,p}$ (for

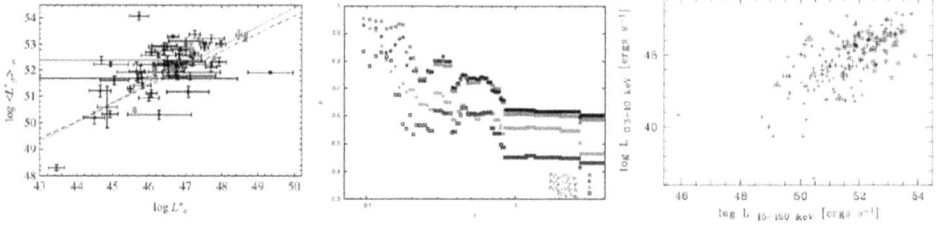

Figure 7.12. Left panel: the log $\langle L_{\gamma,\mathrm{iso}}\rangle_{45}$ – log $L_{X,a}$ correlation (log L^*_p and log L^*_a, respectively, in the plot) for the E4 sample (back points). The black dashed line presents the best-fit, while the E0095 sample (red points) best-fit is indicated by the red line. (Reproduced from Dainotti *et al* (2011b). By permission of Oxford University Press on behalf of the Royal Astronomical Society.) Middle panel: the correlation coefficients for a few correlations versus $u = \sigma_E$ for LGRBs subsets. The log $L_{X,a}$ – log $\langle L_{\gamma,\mathrm{iso}}\rangle_{45}$ correlation is indicated by red squares, the log $L_{X,a}$ – log $\langle L_{\gamma,\mathrm{iso}}\rangle_{90}$ correlation is marked by black circles, the log $L_{X,a}$ – log $\langle L_{\gamma,\mathrm{iso}}\rangle_{T_{X,p}}$ correlation is presented by green asterisks and the log $L_{X,a}$ – log $E_{\gamma,\mathrm{prompt}}$ correlation is displayed by blue squares. (Reproduced from Dainotti *et al* (2011b). By permission of Oxford University Press on behalf of the Royal Astronomical Society.) Right panel: the $L_{X,a}$–$L_{\gamma,\mathrm{iso}}$ correlation with triangles indicating SGRBs and circles marking high-redshift ($z > 3.5$) GRBs. $L_{X,a}$ is computed in the XRT range, while $L_{\gamma,\mathrm{iso}}$ in the BAT range. (Reproduced with permission from Grupe *et al* (2013). Copyright 2013 The American Astronomical Society.)

definitions see section 3.2). The E4 subsample, made up of 62 LGRBs with measured z, and the E0095 subsample, composed of eight GRBs with regular light curves, were employed in the analysis (see black and red points in the left panel of figure 7.12). They concluded that the GRB group providing the highest correlation coefficient, ρ, for the LT correlation also suggests the most significant correlations between prompt–afterglow parameters. Namely, the higher ρ, the smaller the σ_E parameter, as one can see in the middle panel of figure 7.12. In this figure ρ is displayed for the correlations: $(\log\langle L_{\gamma,\mathrm{iso}}\rangle_{45}, \log\langle L_{\gamma,\mathrm{iso}}\rangle_{90}, \log\langle L_{\gamma,\mathrm{iso}}\rangle_{T_{X,p}}, \log E_{\gamma,\mathrm{prompt}}) - \log L_{X,a}$, indicated by red, black, green and blue, respectively. Thus, the increase in the value of ρ can help in choosing the appropriate dataset, which can be used for achieving a more accurate determination of the nature of the plateau phase. Additionally, this set can be used for cosmological purposes. For the IC bursts no tight correlations were found, due to the small amount of data, and it was claimed that their presence does not improve the significance of the correlations.

In conclusion, Dainotti *et al* (2011) stressed the concept that the plateau phase is related to the central engine. In this work, other correlations between log $L_{X,a}$ and parameters of the prompt phase such as log $E_{\gamma,\mathrm{peak}}$ and the variability, log V, were also investigated, recovering (except for V) significant correlations (see table 7.4).

As displayed in table 7.4, the log $T^*_{X,a}$ – log $E_{\gamma,\mathrm{prompt}}$ correlation has a low $\rho = -0.19$. The occurrence of the log $L_{X,a}$ – log$\langle L_{\gamma,\mathrm{iso}}\rangle_{90}$ (see the right panel of figure 7.12) and the log$\langle L_{\gamma,\mathrm{iso}}\rangle_{90}$ – log $T^*_{X,a}$ correlations was also found by Grupe *et al* (2013), who analysed a dataset of 232 GRBs. Given that $\log\langle L_{\gamma,\mathrm{iso}}\rangle_{90} = \log(E_{\gamma,\mathrm{prompt}}/T_{90})$the log$\langle L_{\gamma,\mathrm{iso}}\rangle_{90}$ – log $T^*_{X,a}$ correlation logically follows from the log $T^*_{X,a}$ – log $E_{\gamma,\mathrm{prompt}}$ one.

Table 7.4. In this table are listed: the correlation coefficients ρ, the respective best-fit quantities (a, b), the correlation coefficient r and the occurrence probability P for the prompt–afterglow and prompt–prompt correlations discussed in this section. (Table from Dainotti *et al* (2011b).)

Correlations	E4 ρ	E4 (b, a) P	E0095 ρ	E0095 (b, a) P
$L_{X,a}-\langle L_{\gamma,\mathrm{iso}}\rangle_{45}$	0.59	$(0.67^{+0.14}_{-0.15}, 20.58^{+6.66}_{-6.73})$	0.95	$(0.73^{+0.16}_{-0.11}, 17.90^{+5.29}_{-6.0})$
	0.62	7.7×10^{-8}	0.90	2.3×10^{-3}
$L_{X,a}-\langle L_{\gamma,\mathrm{iso}}\rangle_{90}$	0.60	$(0.63^{+0.15}_{-0.16}, 22.05^{+7.14}_{-7.31})$	0.93	$(0.84^{+0.11}_{-0.12}, 11.86^{+3.43}_{-3.44})$
	0.62	7.7×10^{-8}	0.94	2.7×10^{-3}
$L_{X,\,a}-\langle L_{\gamma,\mathrm{iso}}\rangle_{TX,\,p}$	0.46	$(0.73^{+0.09}_{-0.14}, 16.61^{+4.35}_{-4.35})$	0.95	$(0.93^{+0.20}_{-0.23}, 7.70^{+3.47}_{-3.46})$
	0.56	2.21×10^{-6}	0.90	2.3×10^{-3}
$L_{X,\,a}-E_{\gamma,\mathrm{prompt}}$	0.43	$(0.52^{+0.07}_{-0.06}, 28.03^{+2.98}_{-2.97})$	0.83	$(0.49^{+0.21}_{-0.16}, 29.82^{+7.11}_{-7.82})$
	0.52	1.4×10^{-5}	0.75	3.2×10^{-2}
$T^{*}_{X,a}-E_{\gamma,\mathrm{prompt}}$	-0.19	$(-0.49^{+0.09}_{-0.08}, 54.51^{+0.37}_{-0.30})$	-0.81	$(-0.96^{+0.21}_{-0.22}, 54.67^{+0.69}_{-0.69})$
	-0.21	1.0×10^{-1}	-0.69	5.8×10^{-2}
$L_{X,a}-E_{\gamma,\mathrm{peak}}$	0.54	$(1.06^{+0.53}_{-0.23}, 43.88^{+0.61}_{-1.00})$	0.74	$(1.5^{+0.65}_{-0.94}, 43.10^{+2.53}_{-2.26})$
	0.51	2.2×10^{-5}	0.80	1.7×10^{-2}
$T^{*}_{X,a}-E_{\gamma,\mathrm{peak}}$	-0.36	$(-0.66^{+0.20}_{-0.29}, 4.96^{+0.81}_{-0.80})$	-0.74	$(-1.40^{+0.66}_{-0.65}, 7.04^{+1.79}_{-1.77})$
	-0.35	5.2×10^{-3}	-0.77	2.5×10^{-2}
$\langle L_{\gamma,\mathrm{iso}}\rangle_{45}-E_{\gamma,\mathrm{peak}}$	0.81	$(1.14^{+0.22}_{-0.25}, 49.27^{+0.61}_{-0.60})$	0.76	$(1.45^{+0.26}_{-0.54}, 48.48^{+1.05}_{-1.04})$
	0.67	2.6×10^{-9}	0.92	1.2×10^{-3}

7.5 The $L_{X,a}-L_{X,\mathrm{peak}}$ correlation

Dainotti *et al* (2015) upgraded the analysis of the correlations between prompt and afterglow parameters employing 123 LGRBs detected by *Swift* BAT+XRT with measured z and clear plateau phase. From their work, a correlation between $\log L_{X,\mathrm{peak}}$ and $\log L_{X,a}$ was obtained (for definitions see section 3.2):

$$\log L_{X,a} = A + B \times \log L_{X,\mathrm{peak}}, \qquad (7.5)$$

with $A = -14.67 \pm 3.46$, $B = 1.21^{+0.14}_{-0.13}$, $\rho = 0.79$ and $P < 0.05$ (see the left panel of figure 7.13). $L_{X,\mathrm{peak}}$ is given by:

$$L_{X,\mathrm{peak}} = 4\pi \times D_L(z, \Omega_M, h)^2 \times F_{X,\mathrm{peak}} \times K. \qquad (7.6)$$

ρ of the $\langle \log L_{X,a}-\log L_{\gamma,\mathrm{iso}}\rangle$ correlation (Dainotti *et al* 2011) for 62 LGRBs, namely $\rho = 0.60$, was found to be lower than that of the $\log L_{X,a}-\log L_{X,\mathrm{peak}}$ correlation (see section 7.4). This outcome indicates that a more suitable definition of the luminosity or energy increases ρ by 24%. In the left panel of figure 7.13 $\log L_{X,\mathrm{peak}}$ is computed from the peak flux in X-rays, $F_{X,\mathrm{peak}}$, through a broken or a simple power-law line as the best-fit of the spectrum. This computation has been performed without taking into account the error propagation due to time and energy, and thus it differs from the previously defined luminosities. Given that $\log E_{\gamma,\text{prompt}}$ and $\log E_{\gamma,\text{peak}}$ can be affected by the truncation effect caused by the detector threshold at high and low energies, Dainotti *et al* (2015) preferred the $\log L_{X,a} - \log L_{X,\text{peak}}$ correlation to those described in Dainotti *et al* (2011). Indeed, this truncation issue does not exist for $\log L_{X,\text{peak}}$ (Lloyd and Petrosian 1999). Moreover, the redshift dependence caused by $D_L(z, \Omega_M, h)$ in the $\log L_{X,a} - \log L_{X,\text{peak}}$ correlation was removed by employing fluxes rather than luminosities. The main aim was to test its robustness. This analysis recovered a correlation between $\log F_{X,a}$ and $\log F_{X,\text{peak}}$ with $\rho = 0.63$ (see the right panel of figure 7.13). For a quantitative investigation of the selection effects see chapter 8.

Dainotti *et al* (2015) concluded that the LT correlation and its equivalent in the prompt phase, $\log L_{X,f} - \log T_{X,f}^*$ (Willingale *et al* (2010), see the left panel of figure 7.14) have slopes incompatible at more than 2σ. This result also indicated a difference in the energy and time distributions (see the right panel of figure 7.14). This outcome gave rise to a new interpretation of these observations, see the next section for details.

Later, using 122 LGRBs without XRFs and GRBs associated with SNe, Dainotti *et al* (2016) added a third parameter, $T_{X,a}$, to the $\log L_{X,a} - \log L_{X,\text{peak}}$ correlation, finding

$$\log L_{X,a} = (15.69 \pm 3.8) + (0.67 \pm 0.07) \times \log L_{X,\text{peak}} - (0.80 \pm 0.07)$$
$$\log T_{X,a}, \tag{7.7}$$

with $\rho = 0.93$, $P \leqslant 2.2 \times 10^{-16}$ and $\sigma_{\text{int}} = 0.44 \pm 0.03$. Then, to diminish the scatter of the correlation, they employed only the group of 40 LGRBs with a good number of data and flat plateau phase. These assumptions gave

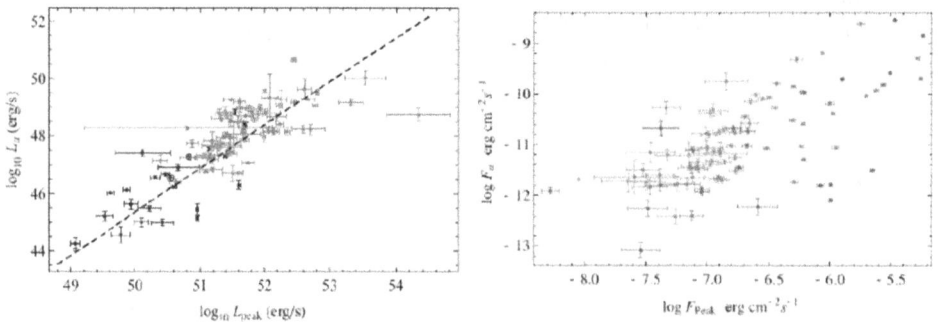

Figure 7.13. Left panel: the $\log L_{X,a} - \log L_{X,\text{peak}}$ correlation with the dataset divided into four different equally populated redshift bins: $z \leqslant 0.8$, shown by blue dots, $0.84 \leqslant z < 1.8$ indicated by magenta dots, $1.8 \leqslant z < 2.9$ represented by green dots and $z \geqslant 2.9$ displayed by red dots. The best-fit line is marked by a dashed line. Right panel: the $\log F_{X,a} - \log F_{X,\text{peak}}$ correlation with the sample divided as in the left panel. $\log F_{X,\text{peak}}$ is obtained as described in the second BAT catalogue in both panels. (Reproduced from Dainotti *et al* (2015a). By permission of Oxford University Press on behalf of the Royal Astronomical Society.)

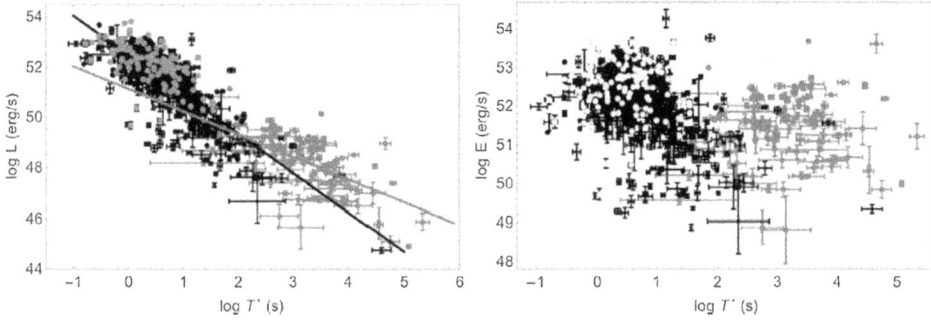

Figure 7.14. The maximum luminosities of the pulses in the prompt phase (log $T_{L\max}$, log $L_{\max}$) are marked by green dots. Right panel: the log E–log T^* correlation for all the prompt and afterglow pulses represented by black and red symbols, respectively. The maximum energy of the pulses in the prompt phase (log $T_{E\max}$, log $E_{\max}$) are displayed by yellow dots. (Reproduced from Dainotti *et al* 2015a. By permission of Oxford University Press on behalf of the Royal Astronomical Society.) For both figures in the prompt pulses log L is log L_{Xf} and log T^* is log T^*_{Xf}, while in the afterglow pulses log L is log $L_{X,a}$ and log T^* is log $T^*_{X,a}$.

$$\log L_{X,a} = (15.75 \pm 5.3) + (0.67 \pm 0.1) \times \log L_{X,\text{peak}} - (0.77 \pm 0.1)\log T_{X,a}, \quad (7.8)$$

with $\rho = 0.90$, $P = 4.41 \times 10^{-15}$ and $\sigma_{\text{int}} = 0.27 \pm 0.04$. From these outcomes it was concluded that the plane defined by this correlation can be regarded as a 'fundamental' plane for GRBs. This can be employed for a better understanding of the theoretical framework of the relation between the prompt and the plateau emissions and as an effective cosmological probe.

7.5.1 Physical interpretation of the $L_{X,a}$–$L_{\gamma,\text{iso}}$ and the $L_{X,a}$–$L_{X,\text{peak}}$ correlations

Dainotti *et al* (2015) showed two different values of the slope for the L–T^* and the E–T^* correlations of the prompt and plateau pulses. These values may imply that the prompt pulses could be produced by the internal shocks, while the plateau pulses may be produced by the external shocks. In the case where the plateau phase is due to synchrotron emission from the external shocks, then all the pulses should present similar initial physical conditions. The association between prompt and afterglow phases was examined to understand the GRB physical models by explaining the log $L_{X,a}$–log $L_{\gamma,\text{iso}}$, the log $L_{X,a}$–log $L_{X,\text{peak}}$ and the LT correlations. The best framework describing this association is that by Hascoët *et al* (2014), which analysed two scenarios. The first, in the FS model, assumes modified microphysics quantities to reduce the initial efficiency of the GRB phenomenon. In the second scenario, the early afterglow phase is produced from a long-lived RS in the FS scenario. In the first case, a wind external medium is presumed together with the internal energy transferred into electrons (or positrons), $\epsilon_e \propto n^{-\nu}$, with n the density medium. When $\nu \approx 1$ a flat plateau phase is obtained. This showed that it is possible to obtain a plateau phase satisfying the log $L_{X,a}$–log $L_{\gamma,\text{iso}}$ and the log $L_{X,a}$–log $L_{X,\text{peak}}$ correlations, even by modifying only one parameter. In the second case, the correlations

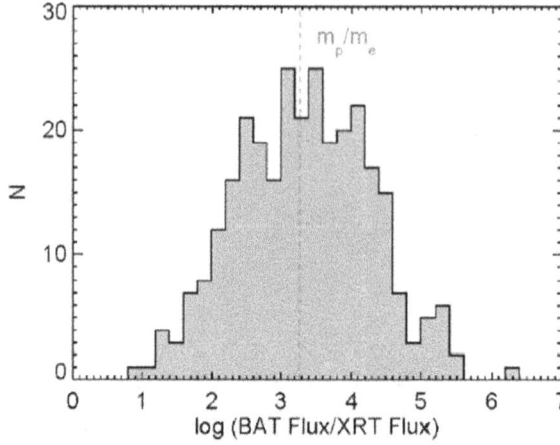

Figure 7.15. Distribution of the ratio of the BAT to XRT fluxes considering a set of *Swift* GRBs. The mean value of this distribution is $\sim 10^3$–10^4. The ratio of the proton to electron masses m_p/m_e is indicated by the vertical line. (Reproduced from Kazanas *et al* (2015). Copyright 2015 Proceedings of Science CC BY 4.0.)

between prompt–afterglow quantities are recovered when Γ of the jet increases with the energy of the burst.

On the other hand, Ruffini *et al* (2014) found that the $\log L_{X,a} - \log L_{\gamma,\mathrm{iso}}$ and the $\log L_{X,a} - \log L_{X,\mathrm{peak}}$ correlations can be explained by the induced gravitational collapse paradigm taking into account the energetic (10^{52}–10^{54} erg) LGRBs with the observed associated SNe. In this case, the external medium fits the plateau and the prompt emissions of the light curves well. Indeed, for the external medium, either a radial structure of the wind (Guida *et al* 2008, Bernardini *et al* 2006, 2007, Caito *et al* 2009) or a division of the shell (Dainotti *et al* 2007) was supposed.

Also the supercritical pile GRB model seems to explain the $\log L_{X,a} - \log L_{\gamma,\mathrm{iso}}$ and the $\log L_{X,a} - \log L_{X,\mathrm{peak}}$ correlations. Indeed, Kazanas *et al* (2015), using this model, concluded that: (1) the ratio, R, between the luminosities is in agreement with the one between the mean prompt flux in the BAT range and the plateau flux in the XRT band, and (2) the ratio, R, has value analogous to that of the proton to electron mass ratio (see figure 7.15). Thus, they pointed out that the $\log L_{X,a} - \log L_{\gamma,\mathrm{iso}}$ and the $\log L_{X,a} - \log L_{X,\mathrm{peak}}$ correlations can be obtained. In conclusion, this appears to be a demanding task for those who aim to better describe the phenomenology of the correlations, taking into account all the aspects mentioned above.

7.6 The $L^F_{O,\mathrm{peak}} - T^{*F}_{O,\mathrm{peak}}$ correlation and its physical interpretation

The correlation between the width, w, and $T_{O,\mathrm{peak}}$ of the light curve flares, expressed by the superscript F, was investigated by Liang *et al* (2010) (for definitions see section 3.2). They employed 32 GRBs detected by *Swift* (see the left panel of figure 7.16). The correlation obtained is the following:

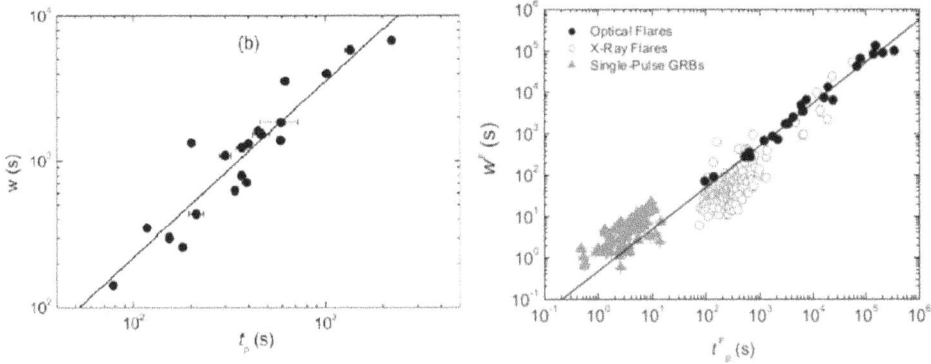

Figure 7.16. Left panel: the $\log w^F - \log T^F_{O,\text{peak}}$ correlation. (Reproduced with permission from Liang *et al* (2010). Copyright 2010 The American Astronomical Society.) Right panel: the $\log w^F - \log T^F_{O,\text{peak}}$ correlation. Both panels indicate the best-fit by a line. (Reproduced with permission from Li *et al* (2012). Copyright 2010 The American Astronomical Society.)

$$\log w^F = (0.05 \pm 0.27) + (1.16 \pm 0.10) \times \log T^F_{O,\text{peak}}, \tag{7.9}$$

with $\rho = 0.94$.

Then, employing 24 flares from 19 single-pulse *CGRO*/BATSE GRBs, Li *et al* (2012) examined a similar correlation, but with smaller values of normalization and slope (see the right panel of figure 7.16). Employing only the 14 GRBs with clear flare activity, they obtained

$$\log w^F = -0.32 + 1.01 \times \log T^F_{O,\text{peak}}. \tag{7.10}$$

From this correlation it was found that flares at earlier times are more luminous and narrower than those at later times. In addition, they claimed that the X-ray flares observed by *Swift*/XRT and optical flares observed in the R filter showed an analogous behaviour with regard to the $w^F - T^F_{O,\text{peak}}$ correlation (Chincarini *et al* 2007, Margutti *et al* 2010), see the right panel of figure 7.16. In their analysis, a new correlation in the rest-frame was also recovered between $L_{O,\text{peak}}$ in the R filter, in units of 10^{48} erg s^{-1}, and $T^*_{O,\text{peak}}$ of the flares for a sample of 19 GRBs (see figure 7.17). The best-fit of this correlation is established by

$$\log L^F_{O,\text{peak}} = (1.89 \pm 0.52) - (1.15 \pm 0.15) \times \log T^{*F}_{O,\text{peak}}, \tag{7.11}$$

with $\rho = 0.85$ and $P < 10^{-4}$. $T^F_{O,\text{peak}}$ was found in the range between ~tens of seconds and ~10^6 seconds, while $L^F_{O,\text{peak}}$ was in the interval between 10^{43} and 10^{49} erg s^{-1}, with a mean of 10^{46} erg s^{-1}. Furthermore, employing only the brightest GRBs of the dataset, they concluded that $T^{*F}_{O,\text{peak}}$ was significantly anti-correlated with $E_{\gamma,\text{prompt}}$ in the range between 1 and 10^4 keV:

$$\log T^{*F}_{O,\text{peak}} = (5.38 \pm 0.30) - (0.78 \pm 0.09) \times \log E_{\gamma,\text{prompt}} / 10^{50}, \tag{7.12}$$

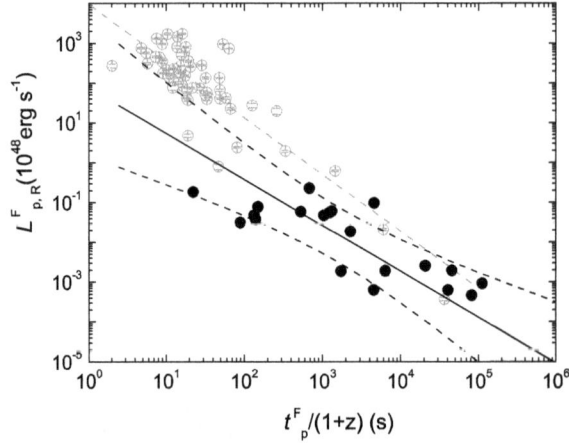

Figure 7.17. The log $L^F_{O,\text{peak}}$–log $T^{*F}_{O,\text{peak}}$ correlation with optical flares represented by black dots, and X-ray flares related to the optical flares displayed by grey circles. Best-fits are indicated by lines. (Reproduced with permission from Li *et al* (2012). Copyright 2012 The American Astronomical Society.)

Table 7.5. Summary of the correlations presented in this section. The correlation in log scale is listed in the first column, the authors are presented in the second and the size of the dataset is displayed in the third. The correlation slope and normalization are indicated in the fourth and fifth columns, while the correlation coefficient and the chance probability, *P*, are shown in the last two columns. (Table from Dainotti and Del Vecchio (2017).)

Correlations	Author	N	Slope	Norm	Corr. coeff.	P
w^F–$T^F_{O,\text{peak}}$	Liang *et al* (2010)	32	$1.16^{+0.10}_{-0.10}$	$0.05^{+0.27}_{-0.27}$	0.94	$<10^{-4}$
	Li *et al* (2012)	19	1.01	−0.32		
$L^F_{O,\text{peak}}$–$T^{*F}_{O,\text{peak}}$	Li *et al* (2012)	19	$-1.15^{+0.15}_{-0.15}$	$1.89^{+0.52}_{-0.52}$	0.85	$<10^{-4}$
$T^{*F}_{O,\text{peak}}$–$E_{\gamma,\text{prompt}}$	Li *et al* (2012)	19	$-0.78^{+0.09}_{-0.09}$	$5.38^{+0.30}_{-0.30}$	0.92	$<10^{-4}$

with $\rho = 0.92$. The correlations mentioned in this section suggest that in the optical band the GRB flares which are more energetic (high values of $E_{\gamma,\text{prompt}}$) display a peak at earlier times and are much brighter. In table 7.5 the correlations presented in this section are shown.

For the physical interpretation of the $L^F_{O,\text{peak}}$–$T^{*F}_{O,\text{peak}}$ correlation, Li *et al* (2012) claimed that the flares are distinct from the afterglow phase. In addition, they concluded that the association between $L^F_{O,\text{peak}}$ and $T^{*F}_{O,\text{peak}}$ implies a common mechanism for the prompt γ-ray and late optical flare radiations, i.e. a central engine repeatedly emitting shells during radiation. Magnetic turbulent reconnection or internal shocks, which possibly stem from the variability (Kobayashi *et al* 1997, Zhang and Yan 2011), could be produced from the collisions among the shells. However, Fenimore *et al* (1995) pointed out that no relevant pattern for the intensity and width of the flare is found by employing GRBs only in the BATSE gamma-ray range. Later, Maxham and Zhang (2009) concluded that a central engine emitting thicker and dimmer shells is needed at late times to understand the w^F–$T^F_{O,\text{peak}}$ correlation. As shown by Perna *et al* (2006) and Proga and Zhang (2006), this could be described with flares produced by blobs, such that the diffusion during the accretion would prolong the accretion time onto the BH.

References

Amati L, Della Valle M, Frontera F, Malesani D, Guidorzi C, Montanari E and Pian E 2007 On the consistency of peculiar GRBs 060218 and 060614 with the $E_{p,i}$–E_{iso} correlation *Astron. Astrophys.* **463** 913–9

Berger E 2007 The prompt gamma-ray and afterglow energies of short-duration gamma-ray bursts *Astrophys. J.* **670** 1254–9

Berger E 2014 Short-duration gamma-ray bursts *Annu. Rev. Astron. Astrophys.* **52** 43–105

Bernardini M G, Bianco C L, Caito L, Chardonnet P, Corsi A, Dainotti M G, Fraschetti F, Guida R, Ruffini R and Xue S S 2006 GRB970228 as a prototype for short GRBs with afterglow *Nuovo Cim.* B **121** 1439–40

Bernardini M G, Bianco C L, Caito L, Dainotti M G, Guida R and Ruffini R 2007 GRB 970228 and a class of GRBs with an initial spikelike emission *Astron. Astrophys.* **474** L13–6

Bernardini M G, Margutti R, Zaninoni E and Chincarini G 2012 A universal scaling for short and long gamma-ray bursts: $E_{X,\text{iso}}$–$E_{3,\text{iso}}$–E_{pk} *Mon. Not. R. Astron. Soc.* **425** 1199–204

Caito L, Bernardini M G, Bianco C L, Dainotti M G, Guida R and Ruffini R 2009 GRB060614: a 'fake' short GRB from a merging binary system *Astron. Astrophys.* **498** 501–7

Chincarini G *et al* 2007 The first survey of x-ray flares from gamma-ray bursts observed by *Swift*: temporal properties and morphology *Astrophys. J.* **671** 1903–20

Dainotti M, Petrosian V, Willingale R, O'Brien P, Ostrowski M and Nagataki S 2015 Luminosity–time and luminosity–luminosity correlations for GRB prompt and afterglow plateau emissions *Mon. Not. R. Astron. Soc.* **451** 3898–908

Dainotti M G and Del Vecchio R 2017 Gamma ray burst afterglow and prompt–afterglow relations: an overview *New Astron. Rev.* **77** 23–61

Dainotti M G, Bernardini M G, Bianco C L, Caito L, Guida R and Ruffini R 2007 GRB 060218 and GRBs associated with supernovae Ib/c *Astron. Astrophys.* **471** L29–32

Dainotti M G, Ostrowski M and Willingale R 2011 Towards a standard gamma-ray burst: tight correlations between the prompt and the afterglow plateau phase emission *Mon. Not. R. Astron. Soc.* **418** 2202–6

Dainotti M G, Postnikov S, Hernandez X and Ostrowski M 2016 A fundamental plane for long gamma-ray bursts with x-ray plateaus *Astrophys. J.* **825** L20

D'Avanzo P *et al* 2012 A complete sample of bright *Swift* gamma-ray bursts: x-ray afterglow luminosity and its correlation with the prompt emission *Mon. Not. R. Astron. Soc.* **425** 506–13

Fenimore E E, in't Zand J J M, Norris J P, Bonnell J T and Nemiroff R J 1995 Gamma-ray burst peak duration as a function of energy *Astrophys. J.* **448** L101

Gehrels N *et al* 2008 Correlations of prompt and afterglow emission in *Swift* long and short gamma-ray bursts *Astrophys. J.* **689** 1161–72

Gehrels N 2007 *Short GRB Prompt and Afterglow Correlations* https://ntrs.nasa.gov/archive/nasa/casi.ntrs.nasa.gov/20080045485.pdf.

Ghisellini G, Nardini M, Ghirlanda G and Celotti A 2009 A unifying view of gamma-ray burst afterglows *Mon. Not. R. Astron. Soc.* **393** 253–71

Grupe D, Nousek J A, Veres P, Zhang B-B and Gehrels N 2013 Evidence for new relations between gamma-ray burst prompt and x-ray afterglow emission from 9 years of *Swift Astrophys. J. Suppl. Ser.* **209** 20

Guida R, Bernardini M G, Bianco C L, Caito L, Dainotti M G and Ruffini R 2008 The Amati relation in the 'fireshell' model *Astron. Astrophys.* **487** L37–40

Hascoët R, Daigne F and Mochkovitch R 2014 The prompt–early afterglow connection in gamma-ray bursts: implications for the early afterglow physics *Mon. Not. R. Astron. Soc.* **442** 20–7

Jakobsson P, Hjorth J, Fynbo J P U, Watson D, Pedersen K, Björnsson G and Gorosabel J 2004 *Swift* identification of dark gamma-ray bursts *Astrophys. J.* **617** L21–4

Kaneko Y *et al* 2007 Prompt and afterglow emission properties of gamma-ray bursts with spectroscopically identified supernovae *Astrophys. J.* **654** 385–402

Kann D A *et al* 2010 The afterglows of *Swift*-era gamma-ray bursts. I. Comparing pre-*Swift* and *Swift*-era long/soft (Type II) GRB optical afterglows *Astrophys. J.* **720** 1513–58

Kazanas D, Racusin J L, Sultana J and Mastichiadis A 2015 The statistics of the prompt-to-afterglow GRB flux ratios and the supercritical pile GRB model, arXiv: 1501.01221

Kobayashi S, Piran T and Sari R 1997 Can internal shocks produce the variability in gamma-ray bursts? *Astrophys. J.* **490** 92

Li L *et al* 2012 A comprehensive study of gamma-ray burst optical emission. I. Flares and early shallow-decay component *Astrophys. J.* **758** 27

Liang E-W, Zhang B-B and Zhang B 2007 A comprehensive analysis of *Swift* XRT data. II. Diverse physical origins of the shallow decay segment *Astrophys. J.* **670** 565–83

Liang E-W, Yi S-X, Zhang J, Lü H-J, Zhang B-B and Zhang B 2010 Constraining gamma-ray burst initial Lorentz factor with the afterglow onset feature and discovery of a tight Γ_0–$E_{\mathrm{gamma,iso}}$ correlation *Astrophys. J.* **725** 2209–24

Lloyd N M and Petrosian V 1999 Distribution of spectral characteristics and the cosmological evolution of gamma-ray bursts *Astrophys. J.* **511** 550–61

Margutti R, Guidorzi C, Chincarini G, Bernardini M G, Genet F, Mao J and Pasotti F 2010 Lag–luminosity relation in γ-ray burst x-ray flares: a direct link to the prompt emission *Mon. Not. R. Astron. Soc.* **406** 2149–67

Margutti R *et al* 2013 The prompt–afterglow connection in gamma-ray bursts: a comprehensive statistical analysis of *Swift* x-ray light curves *Mon. Not. R. Astron. Soc.* **428** 729–42

Maxham A and Zhang B 2009 Modeling gamma-ray burst x-ray flares within the internal shock model *Astrophys. J.* **707** 1623–33

Nardini M, Ghisellini G, Ghirlanda G, Tavecchio F, Firmani C and Lazzati D 2006 Clustering of the optical-afterglow luminosities of long gamma-ray bursts *Astron. Astrophys.* **451** 821–33

Nousek J A *et al* 2006 Evidence for a canonical gamma-ray burst afterglow light curve in the *Swift* XRT data *Astrophys. J.* **642** 389–400

Nysewander M, Fruchter A S and Pe'er A 2009 A comparison of the afterglows of short- and long-duration gamma-ray bursts *Astrophys. J.* **701** 824–36

Oates S R, Racusin J L, De Pasquale M, Page M J, Castro-Tirado A J, Gorosabel J, Smith P J, Breeveld A A and Kuin N P M 2015 Exploring the canonical behaviour of long gamma-ray bursts using an intrinsic multiwavelength afterglow correlation *Mon. Not. R. Astron. Soc.* **453** 4121–35

Panaitescu A and Vestrand W T 2011 Optical afterglows of gamma-ray bursts: peaks, plateaus and possibilities *Mon. Not. R. Astron. Soc.* **414** 3537–46

Perna R, Armitage P J and Zhang B 2006 Flares in long and short gamma-ray bursts: a common origin in a hyperaccreting accretion disk *Astrophys. J.* **636** L29–32

Proga D and Zhang B 2006 The late time evolution of gamma-ray bursts: ending hyperaccretion and producing flares *Mon. Not. R. Astron. Soc.* **370** L61–5

Racusin J L *et al* 2011 *Fermi* and *Swift* gamma-ray burst afterglow population studies *Astrophys. J.* **738** 138

Rowlinson A, O'Brien P T, Metzger B D, Tanvir N R and Levan A J 2013 Signatures of magnetar central engines in short GRB light curves *Mon. Not. R. Astron. Soc.* **430** 1061–87

Ruffini R, Muccino M, Bianco C L, Enderli M, Izzo L, Kovacevic M, Penacchioni A V, Pisani G B, Rueda J A and Wang Y 2014 On binary-driven hypernovae and their nested late x-ray emission *Astron. Astrophys.* **565** L10

Willingale R *et al* 2007 Testing the standard fireball model of gamma-ray bursts using late x-ray afterglows measured by *Swift Astrophys. J.* **662** 1093–110

Willingale R, Genet F, Granot J and O'Brien P T 2010 The spectral–temporal properties of the prompt pulses and rapid decay phase of gamma-ray bursts *Mon. Not. R. Astron. Soc.* **403** 1296–316

Zhang B and Kobayashi S 2005 Gamma-ray burst early afterglows: reverse shock emission from an arbitrarily magnetized ejecta *Astrophys. J.* **628** 315–34

Zhang B and Yan H 2011 The internal-collision-induced magnetic reconnection and turbulence (ICMART) model of gamma-ray bursts *Astrophys. J.* **726** 90

Zhang B *et al* 2007a GRB radiative efficiencies derived from the *Swift* data: GRBs versus XRFs, long versus short *Astrophys. J.* **655** 989–1001

Chapter 8

Selection effects in the afterglow and prompt–afterglow correlations

GRB correlations are often influenced by selection effects. Selection effects are distortions or biases that appear when the considered sample does not fully characterize the 'true' distribution. It was stressed by Efron and Petrosian (1992), Lloyd and Petrosian (1999), Dainotti *et al* (2013a, 2015b) and Petrosian *et al* (2013) that in the analysis of a multivariate sample a significant point is to recover the correlations without selection biases (intrinsic correlations) and to compute the correct distribution for each parameter.

The correlations mentioned in previous sections are mostly affected by two types of biases: the first caused by the redshift dependence of the parameters and the latter originating from the threshold of the detector employed. Several methods to deal with selection biases are presented here. In section 8.1, the redshift induced correlations are presented through a qualitative procedure, in section 8.2 a more quantitative procedure through the EP method is discussed. In section 8.3, how to obtain the intrinsic correlations corrected for selection biases is described. In section 8.4 the selection biases for the optical and X-ray luminosities are summarized. Finally, in section 8.5 the estimation of the intrinsic correlation through Monte Carlo simulations is displayed.

8.1 Redshift induced correlations

Given that a modification of the correlation slope, b, of the LT correlation has been pointed out by different analyses (Dainotti *et al* 2008, 2010), Dainotti *et al* (2011) investigated the redshift dependence of the LT correlation to understand the motivation of this variation. Indeed, even if the values are compatible within 1 σ, Dainotti *et al* (2008) obtained as the central value $b = -0.74^{+0.20}_{-0.19}$, while Dainotti *et al* (2010) computed $b = -1.06^{+0.27}_{-0.28}$. Although the 62 LGRBs of the dataset were found to not be equally distributed in the interval $(z_{min}, z_{max}) = (0.08, 8.26)$, the main point was to examine if the coefficients (a, b, σ_{int}) were compatible within the errors in the whole redshift range (see the left panel of figure 8.1).

Consequently, the sample was divided into three redshift groups with equal numbers of GRBs, Z1 = (0.08, 1.56), Z2 = (1.71, 3.08) and Z3 = (3.21, 8.26), displayed as the blue, green and red points, respectively, in the left panel of figure 8.1. Table 8.1 shows the outcomes.

The computed ρ was significant in each redshift bin, corroborating the redshift independence of the LT correlation. Indeed, if ρ was not significant in one or more redshift bins, this would have implied that the correlation as a whole would have been reproduced merely as an effect of the redshift dependence. As is visible from table 8.1, b for groups Z1 and Z2 are compatible at the 68% CL, while b for groups Z1 and Z3 only at the 95% CL. On the other hand, the normalization a is similar in all the groups. Given the paucity of GRBs in the sample and the existence of high σ_E GRBs, it could not be verified if the LT correlation is shallower for higher z GRBs.

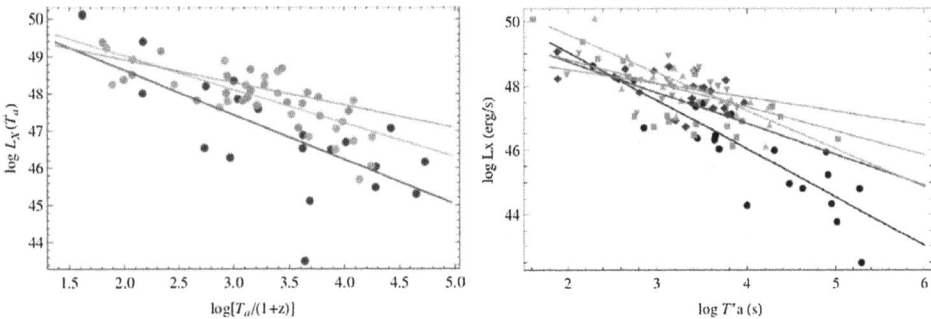

Figure 8.1. Left panel: the LT correlation with the dataset split into three redshift ranges: Z1 = (0.08, 1.56) (blue dots), Z2 = (1.71, 3.08) (green dots) and Z3 = (3.21, 8.26) (red dots). The best-fit lines are drawn with the same colours. (Reproduced with permission from Dainotti *et al* (2011a). Copyright 2011 The American Astronomical Society.) Right panel: the LT correlation using 101 GRBs split into five redshift intervals: $z < 0.89$ (black dots), $0.89 \leqslant z \leqslant 1.68$ (magenta dots), $1.68 < z \leqslant 2.45$ (blue dots), $2.45 < z \leqslant 3.45$ (green dots) and $z \geqslant 3.45$ (red dots). The best-fits are indicated by solid lines. (Reproduced with permission from Dainotti *et al* (2013a). Copyright 2013 The American Astronomical Society.)

Table 8.1. Calibration results for GRBs split into three equally populated redshift bins. The Z1, Z2 and Z3 bins are in the following ranges, respectively: $(z_{min}, z_{max}) = (0.08, 1.56), (1.71, 3.08), (3.21, 8.26)$. The best-fit values are indicated by the subscript 'bf', while the median values are represented by the *median* subscript. (Table from Dainotti *et al* (2011a).)

Id	ρ	$(b, a, \sigma_{int})_{bf}$	b_{median}	$(\sigma_{int})_{median}$
Z1	−0.69	(−1.20, 51.04, 0.98)	$-1.08^{+0.27}_{-0.30}$	$1.01^{+0.20}_{-0.16}$
Z2	−0.83	(−0.90, 50.82, 0.43)	$-0.86^{+0.18}_{-0.16}$	$0.45^{+0.09}_{-0.08}$
Z3	−0.63	(−0.61, 50.14, 0.26)	$-0.58^{+0.14}_{-0.15}$	$0.26^{+0.07}_{-0.06}$

Therefore, it was concluded that broader datasets with low σ_E and a more regular binning of z are needed to solve this issue.

Using 101 GRBs, Dainotti *et al* (2013a) analysed the updated sample dividing it into five redshift ranges with equal numbers of GRBs, thus having 20 GRBs in each redshift range. The bins are displayed in the right panel of figure 8.1 by distinct colours: black for $z < 0.89$, magenta for $0.89 \leqslant z \leqslant 1.68$, blue for $1.68 < z \leqslant 2.45$, green for $2.45 < z \leqslant 3.45$ and red for $z \geqslant 3.45$. The best-fit line for each redshift bin is presented with the same colours. In each bin b is distinct for the different samples. In addition, GRBs in each redshift range in the different bins are well separated in the LT plane. In the left panel of figure 8.2 the evolution of the slope of the LT correlation is shown for each redshift range as a function of the average redshift of each bin. Later, Dainotti *et al* (2015b, 2015a) split the dataset of 176 GRBs into five redshift bins, as displayed in the right panel of figure 8.2. We found a weak dependence on z given by: $b(z) = 0.10z - 1.38$.

Furthermore, dividing the dataset into four redshift bins (see the left panel of figure 7.13), Dainotti *et al* (2015b) concluded that the $\log L_{X,a} - \log L_{X,\text{peak}}$ correlation is not dependent on the redshift. Indeed, the distribution of the GRBs is not clustered around a particular area, suggesting no significant redshift evolution. Regarding the luminosity, a weak redshift evolution for $\log L_{X,a}$ was claimed by Dainotti *et al* (2013a), and instead an important redshift evolution for $\log L_{X,\text{peak}}$ was computed by Yonetoku *et al* (2004), Petrosian *et al* (2013) and Dainotti *et al* (2015b). For more details, see sections 8.2.1 and 8.2.2.

8.2 Redshift induced correlations through the Efron and Petrosian method

To examine quantitatively the dependence on the redshift of the physical quantities and to remove the selection effects, the EP method was employed for the analysis of incomplete GRB samples (Petrosian *et al* 2009, Lloyd and Petrosian 1999, Lloyd *et al* 2000). This method calculates the redshift evolution and the instrumental

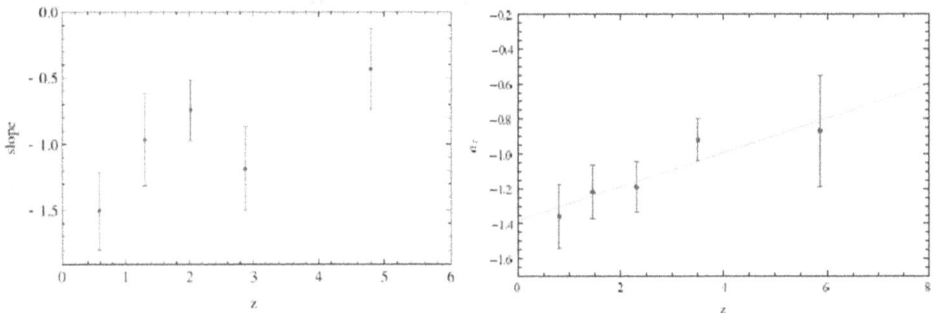

Figure 8.2. Left panel: the slope b as a function of the average value of the redshift intervals. (Reproduced with permission from Dainotti *et al* (2013a). Copyright 2013 The American Astronomical Society.) Right panel: α_τ, corresponding to the slope b, as a function of z. The best-fit function is given by $\alpha_\tau = 0.10z - 1.38$. (Reproduced with permission from Dainotti *et al* (2015b, 2015a). Copyright 2015 The American Astronomical Society.)

selection biases with the aim of overcoming them. The final goal of using this method is to allow the computation of the intrinsic correlation slope b_{int} by defining new bias-free quantities, the so-called local variables, indicated with the symbol $'$. Namely, these are the variables as they are computed at $z = 0$ in the local universe. Then, the EP method employs a variation to the Kendall tau test, τ, to calculate the most reliable estimates of the parameters in the functions describing the luminosity and time evolution. For details on the meaning and determination of τ see Efron and Petrosian (1992).

8.2.1 Luminosity evolution

To investigate the relation between luminosity and redshift (the so-called luminosity evolution) for both prompt and plateau phases, the limiting plateau flux, F_{lim}, should be parameterized before employing the EP method. F_{lim} is the smallest observed flux at a certain z. The limit given by XRT, $F_{lim,XRT} = 10^{-14}$ erg cm^{-2} s^{-1}, does not describe the sample incompleteness well. The best evaluation for the flux threshold is given by Cannizzo $et\ al$ (2011) (10^{-12} erg cm^{-2} s^{-1}). Between several threshold fluxes, Dainotti $et\ al$ (2013a) chose the value $F_{lim,XRT} = 1.5 \times 10^{-12}$ erg cm^{-2} s^{-1}, which allowed for the inclusion of 90 GRBs (the initial sample was composed of 101 GRBs, see the left panel of figure 8.3). Instead, for the prompt phase, Dainotti $et\ al$ (2015b) computed a prompt limiting flux $F_{lim,BAT} = 4 \times 10^{-8}$ erg cm^{-2} s^{-1}, which also included 90% of the data sample (see the right panel of figure 8.3).

From the analysis by Dainotti $et\ al$ (2013a), was chosen a correlation function, $g(z)$, such that $L'_{X,a} \equiv L_{X,a}/g(z)$ does not depend on z. This function is simply given by:

$$g(z) = (1 + z)^{k_{L_{X,a}}}. \tag{8.1}$$

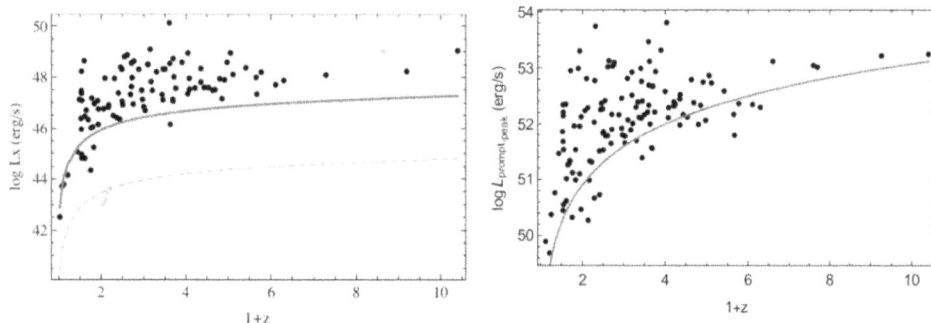

Figure 8.3. Left panel: the $\log L_{X,a}$–z distribution with the XRT flux limit, 1.0×10^{-14} erg cm^{-2} s^{-1}, represented by a dashed green line. This value does not describe the limit of the flux well. The best flux limit for the considered dataset, 1.5×10^{-12} erg cm^{-2} s^{-1}, is indicated by a solid red line. (Reproduced with permission from Dainotti $et\ al$ (2013a). Copyright 2013 The American Astronomical Society.) Right panel: the $\log L_{X,peak}$–z distribution, with the BAT flux limit (4.0×10^{-8} erg cm^{-2} s^{-1}) and K correction $K = 1$, describes the flux limit for the considered dataset well, displayed by a solid red line. (Reproduced from Dainotti $et\ al$ (2015b). By permission of Oxford University Press on behalf of the Royal Astronomical Society.)

More complicated functional forms were also considered by Dainotti *et al* (2013a, 2015b), although they provided similar outcomes.

From this analysis, Dainotti *et al* (2013a) computed the value of $k_{L_{X,a}}$ for which $\tau_{L_{X,a}} = 0$. This is the best definition of the luminosity evolution at the 1 σ level. They calculated $k_{L_{X,a}} = -0.05^{+0.35}_{-0.55}$, indicating a weak evolution (see the left panel of figure 8.4). In the same picture, using the dataset of 47 GRBs (green dotted line) in common with the sample of 77 LGRBs employed in Dainotti *et al* (2011), the analogous distribution for this smaller sample is displayed. It was claimed that the outcomes of the afterglow luminosity evolution for these two sets are in agreement within 2 σ.

In the prompt phase, for the evolution of $L_{X,\text{peak}}$, Dainotti *et al* (2015b) employed both a simple relation function (see equation (8.1)) and a more complex one:

$$g(z) = \frac{Z^{k_L}(1 + Z^{k_L}_{cr})}{Z^{k_L} + Z^{k_L}_{cr}}, \tag{8.2}$$

where $Z = 1 + z$ and $Z_{cr} = 3.5$. For the prompt phase a significant luminosity evolution was recovered, $k_{L_{X,\text{peak}}} = 2.13^{+0.33}_{-0.37}$, through the simple relation, and $k_{L_{X,\text{peak}}} = 3.09^{+0.40}_{-0.35}$ in the case of the more complex function (see the middle and right panels of figure 8.4). In conclusion, the outcomes of the prompt luminosity evolution for these two functions are in agreement within 2 σ.

8.2.2 Time evolution

As has already been done for the luminosity, the limit of the plateau end time should also be computed. Dainotti *et al* (2013a) obtained $T^*_{X,a,\text{lim}} = 242/(1 + z)$ s, while (Dainotti *et al* 2015b) claimed that for the prompt phase $T^*_{X,\text{prompt,lim}} = 1.74/(1 + z)$ s (see figures 8.5 and 8.6). To analyse the time evolution, Dainotti *et al* (2013a) defined the function $f(z)$, again such that $T'_{X,a} \equiv T^*_{X,a}/f(z)$ is independent of z. This is the equivalent of equation (8.1) in which $k_{LX,a}$ is substituted by the coefficient of the time evolution, $k_{T^*_{X,a}}$. Similarly to the luminosity evolution, for the time evolution the best

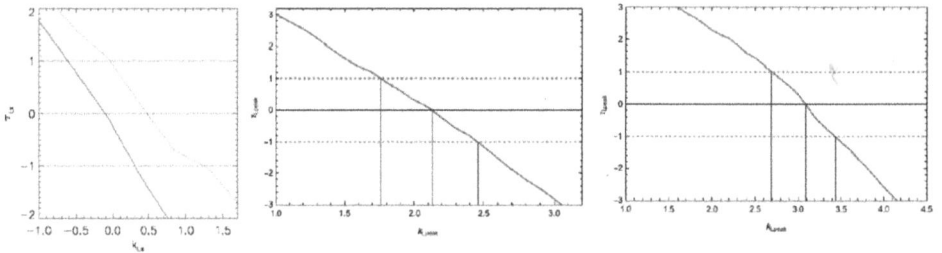

Figure 8.4. Left panel: the τ–$k_{L_{X,a}}$ distribution for the complete dataset shown by a red line. The green dotted line represents the distribution for 47 GRBs in common with the 77 GRB sample used by Dainotti *et al* (2011a). (Reproduced with permission from Dainotti *et al* (2013a). Copyright 2013 The American Astronomical Society.) Middle panel: the τ–$k_{L_{X,\text{peak}}}$ distribution obtained through equation (8.1). Right panel: the τ–$k_{L_{X,\text{peak}}}$ distribution computed through equation (8.2). (Reproduced from Dainotti *et al* 2015b. By permission of Oxford University Press on behalf of the Royal Astronomical Society.)

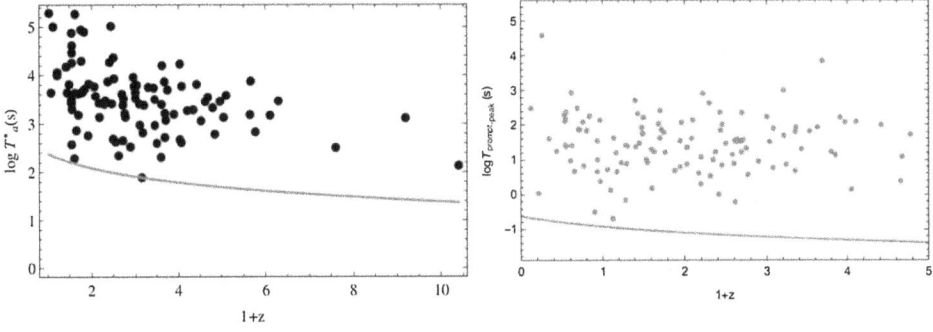

Figure 8.5. Left panel: the $\log T^*_{X,a}$–z distribution with the limiting time in the rest-frame indicated by a red line. For this dataset the limiting end time of the plateau phase in the observed frame is set at $T_{X,a,\mathrm{lim}} = 242$ s. (Reproduced with permission from Dainotti *et al* (2013a). Copyright 2013 The American Astronomical Society.) Right panel: the $\log T^*_{X,\mathrm{prompt}}$–$z$ distribution with the limiting rest-frame time, $\log(T_{X,\mathrm{prompt,lim}}/(1 + z))$, displayed by a red line. $\log T^*_{X,\mathrm{prompt}}$ is the total width of the peak pulses in each GRB pulse. For this dataset the limiting pulse width in the observed frame is set at $\log T_{X,\mathrm{prompt,lim}} = 0.24$ s. (Reproduced from Dainotti *et al* (2015b). By permission of Oxford University Press on behalf of the Royal Astronomical Society.)

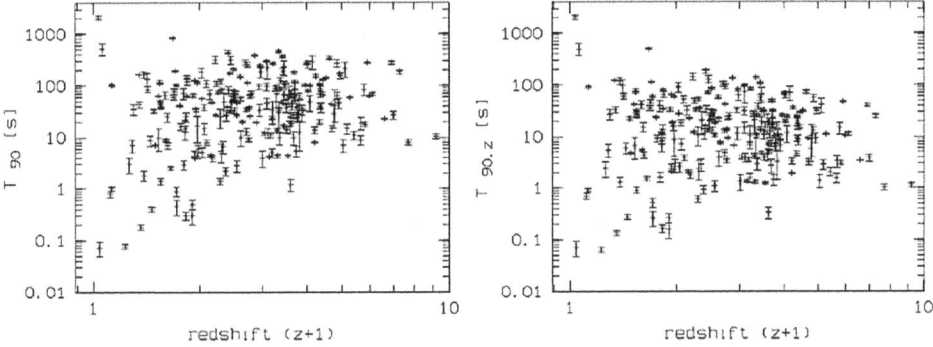

Figure 8.6. T_{90} in the BAT energy band in the observed (displayed in the left panel) and in the rest-frame (displayed in the right panel) versus the redshift. (Reproduced with permission from Grupe *et al* (2013). Copyright 2013 The American Astronomical Society.)

value of $k_{T^*_{X,a}}$ is the one for which $\tau_{T^*_{X,a}} = 0$. As a result, the $\tau_{T^*_{X,a}}$ versus $k_{T^*_{X,a}}$ distribution indicates a significant evolution for $T^*_{X,a}$ ($k_{T^*_{X,a}} = -0.85^{+0.30}_{-0.30}$, see the left panel of figure 8.7). The analogous distribution for a smaller set of 47 GRBs, in common with the 77 GRBs employed in Dainotti *et al* (2011), is displayed in the same picture with a green dotted line. The findings of the afterglow time evolution for these two sets are in agreement within 1.5 σ.

In the analysis of the prompt time evolution, in addition to the simple relation function, Dainotti *et al* (2015b) employed a more complex function. This is the same as equation (8.2), but with $k_{LX,\mathrm{peak}}$ substituted by the coefficient of the time evolution, $k_{T^*_{X,\mathrm{prompt}}}$.

Finally, for both the simple function and the more complex one, a significant time evolution in the prompt was not found ($k_{T^*X,\text{prompt}} = -0.62^{+0.38}_{-0.38}$ and $k_{T^*X,\text{prompt}} = -0.17^{+0.24}_{-0.27}$, respectively, see the middle and right panels of figure 8.7). It was concluded that the results of the prompt time evolution for these two functions agree within 1 σ.

8.3 Evaluation of the intrinsic slope

Computing the 'true' slope of the correlation is the next stage to study an intrinsic correlation. Dainotti *et al* (2013a) considered the EP method in the local time ($T'_{X,a}$) and luminosity ($L'_{X,a}$) space to compute b_{int} for the LT correlation, $b_{\text{int}} = -1.07^{+0.09}_{-0.14}$, at 12 σ CL. This result is clearly visible from the left panel of figure 8.8 (Dainotti *et al* 2013b). Otherwise, in the case of no correlation it would have resulted that $\tau = 0$ for $b_{\text{int}} = 0$ at 1 σ. On the other hand, to compute b_{int} of the log $L_{X,a}$–log $L_{X,\text{peak}}$

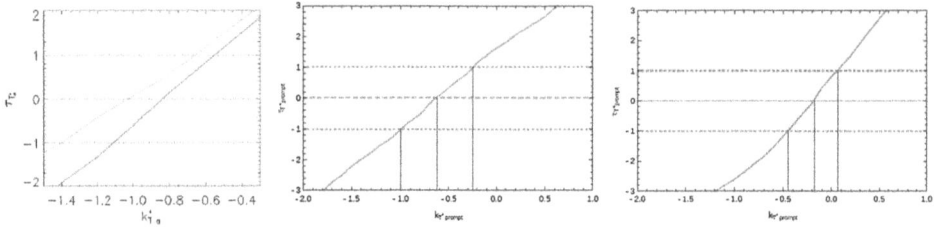

Figure 8.7. Left panel: the τ–$k_{T^*_{X,a}}$ distribution with the complete dataset indicated by a red line. The green dotted line represents 47 GRBs in common with the 77 GRB sample shown in Dainotti *et al* (2011a). (Reproduced with permission from Dainotti *et al* (2013a). Copyright 2013 The American Astronomical Society.) Middle panel: the τ–$k_{T^*_{X,\text{prompt}}}$ distribution obtained through the equivalent equation (8.1) for the time. Right panel: the τ–$k_{T^*_{X,\text{prompt}}}$ distribution computed through the equivalent equation (8.2) for the time. (Reproduced from Dainotti *et al* (2015b). By permission of Oxford University Press on behalf of the Royal Astronomical Society.)

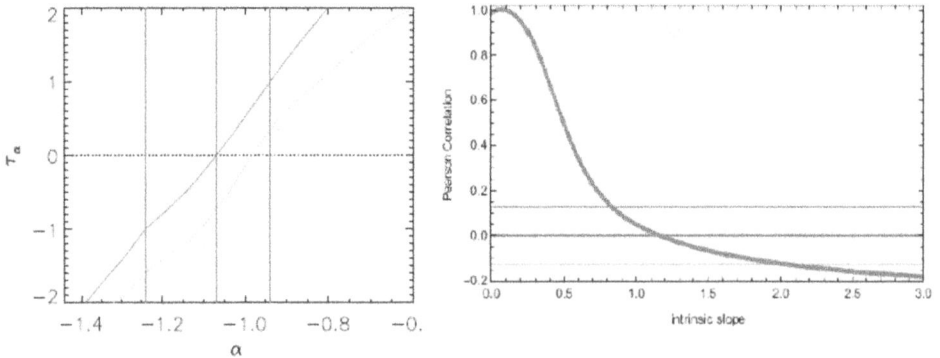

Figure 8.8. Left panel: the τ–b_{int} distribution. In the plot b_{int} is represented by α. (Reproduced with permission from Dainotti *et al* (2013a). Copyright 2013 The American Astronomical Society.) Right panel: the Pearson correlation coefficient, r, versus b_{int} for the log $L_{X,\text{peak}}$–log $L_{X,a}$ correlation with the best-fit indicated by a thick line. The 0.05% probability that the distribution of data is due to chance is represented by two thinner lines. (Reproduced from Dainotti *et al* (2015b). By permission of Oxford University Press on behalf of the Royal Astronomical Society.)

correlation, Dainotti *et al* (2015b) applied the partial correlation coefficient. The partial correlation coefficient represents the level of correlation between two quantities as a function of b_{int} given by:

$$r_{L'_{X,peak}L'_{X,a}, D_L} = \frac{r_{L'_{X,peak}, L'_{X,a}} - r_{L'_{X,peak}, D_L} *r_{L'_{X,a}, D_L}}{\left(1 - r^2_{L'_{X,peak}, D_L}\right)*\left(1 - r^2_{L'_{X,a}, D_L}\right)},$$ (8.3)

where $\log L'_{X,a} = L'_{X,a}$ and $\log L'_{X,peak} = L'_{X,peak}$.

The right panel of figure 8.8 shows that the correlation is particularly relevant when $b_{int} = 1.14^{+0.83}_{-0.32}$, which is within 1 σ of the observed slope, b. In addition, in the same way as Butler *et al* (2010), Dainotti *et al* (2015b, 2015a) mimicked a set with biases on both time and luminosity. These selection effects were similar for every monotonic efficiency function related to the detection of luminosity and time. The final aim was to display how an unknown efficiency function influences b and the density rate of GRBs. In addition, biases in slope or normalization due to truncation were examined, resulting in different estimates. In this way, the scatter of the correlation and its selection effects were also investigated.

In conclusion, this work and that by Dainotti *et al* (2013a) have displayed that the LT correlation can be corrected for selection effects and employed as redshift estimator (see chapter 9) and cosmological tool (see section 5.2). The partial correlation coefficient method was also applied by D'Avanzo *et al* (2012) to the $L_{X,a}$–$E_{\gamma,prompt}$ correlation, by Oates *et al* (2015) to the $L_{O,200s}$–$\alpha_{O,>200s}$ correlation, and by Racusin *et al* (2016) to the $L_{X,200s}$–$\alpha_{X,>200s}$ correlation. Their aim was to prove that these correlations are not affected by the redshift dependence.

8.4 Selection effects for the optical and X-ray luminosities

Here, the selection biases affecting the optical and X-ray luminosities are presented for the correlations discussed so far. Studying the optically dark afterglow phase, Nardini *et al* (2008) analysed if the observed luminosity distribution is the result of selection effects. They simulated the $\log L_{O,12}$, z, the host galaxy dust absorption, A_V^{host}, and the telescope limiting magnitude for the whole sample of 30 000 GRBs to compare the observed optical luminosity distribution to the simulated one. They claimed that only GRBs with a flux greater than the detector threshold flux need to be considered. This is equivalent to a lower luminosity truncation around $\log L_{O,12} \approx 31.2$ (erg s^{-1} Hz^{-1}). GRBs with such a luminosity are not detected, thus constraining the luminosity function. In addition, a group of low-luminosity GRBs which are 3.6 σ away from the mean of the distribution were also investigated from a statistical point of view. They found that if the absorption depends on the frequency, the observed luminosity distribution does not fit with any unimodal one. Nevertheless, a unimodal luminosity distribution can be recovered if the majority of GRBs are absorbed by 'grey' achromatic dust. Thus, an optically subluminous sample, or a group of GRBs for which a significant achromatic absorption is detected, could explain dark bursts.

For the investigation of the biases of $L_{O,\text{peak}}$, considerations on the selection effects of $F_{O,\text{peak}}$ observations are required. Indeed, for a typical optical afterglow ($F_{O,a} \propto T_{O,a}^{-1}$), Panaitescu and Vestrand (2008) claimed that a shallower $\log F_{O,\text{peak}} - \log T_{O,\text{peak}}$ anti-correlation than the measured one can be produced by changes in the observer offset angle. Indeed, the slope of the $\log F_{O,\text{peak}} - \log T_{O,\text{peak}}$ anti-correlation becomes steeper due to an observational selection bias. Gehrels *et al* (2008) found out that *Swift* SGRBs are fluence-limited, while *Swift* LGRBs are flux-limited due to the instrument trigger.

As found by Nysewander *et al* (2009), the absorption of photons in the host galaxy can modify the ratio $F_{O,11}/F_{X,11}$. In addition, they claimed that $F_{X,11}$ should be correct, because the LGRBs detected in the XRT energy range do not show X-ray column absorptions, unlike most LGRBs. The presence of optical absorption (A_V) in LGRB afterglow phases suggests column densities (N_H) with lower values than those in the X-ray, while A_V are roughly one-tenth to one magnitude (Schady *et al* 2007, Cenko *et al* 2009). In contrast, in SGRBs the optical emission relative to the X-ray one is brighter than the one expected from the standard model. Kann *et al* (2010) concluded that the clustering of $L_{O,1d}$ is not significant for *Swift* GRBs, as claimed by Liang and Zhang (2006b) and Nardini *et al* (2006). Thus, the clustering observed in pre-*Swift* data can be caused by the selection effects alone. Finally, Berger (2014) pointed out that the detection of the optical afterglow phase can modify the luminosity distribution favouring regions with a particularly high-density medium.

8.5 Selection effects in the $L_{O,200\text{s}} - \alpha_{O,>200\text{s}}$ correlation

Employing the criteria from Oates *et al* (2009, 2012) allows one to obtain high-S/N light curves at early and late times from the UVOT observations. If the dimmest optical/UV afterglow phases decay more slowly than the most luminous ones, the luminosity distribution at late time is narrower and ρ of the $\log L_{O,200\text{s}} - \alpha_{O,>200\text{s}}$ correlation have to eventually be almost marginal. Given that both of these effects were found in their sample, there was the possibility that the $\log L_{O,200\text{s}} - \alpha_{O,>200\text{s}}$ correlation could be caused by chance due to the choice of the dataset. Therefore, they carried out Monte Carlo simulations to check this possibility. They concluded that, up to 10^6 trials, 34 have a correlation coefficient ρ with higher value than the original one. Their result implied that at 4.2 σ CL the selection criteria do not affect the $\log L_{O,200\text{s}} - \alpha_{O,>200\text{s}}$ correlation, and that this correlation is not due to chance.

References

Berger E 2014 Short-duration gamma-ray bursts *Annu. Rev. Astron. Astrophys.* **52** 43–105

Butler N R, Bloom J S and Poznanski D 2010 The cosmic rate, luminosity function, and intrinsic correlations of long gamma-ray bursts *Astrophys. J.* **711** 495–516

Cannizzo J K, Troja E and Gehrels N 2011 Fall-back disks in long and short gamma-ray bursts *Astrophys. J.* **734** 35

Cenko S B *et al* 2009 Dark bursts in the *Swift* era: the Palomar 60 inch-*Swift* early optical afterglow catalog *Astrophys. J.* **693** 1484–93

Dainotti M G, Cardone V F and Capozziello S 2008 A time–luminosity correlation for γ-ray bursts in the x-rays *Mon. Not. R. Astron. Soc.* **391** L79–83

Dainotti M G, Willingale R, Capozziello S, Fabrizio Cardone V and Ostrowski M 2010 Discovery of a tight correlation for gamma-ray burst afterglows with 'canonical' light curves *Astrophys. J.* **722** L215–9

Dainotti M G, Fabrizio Cardone V, Capozziello S, Ostrowski M and Willingale R 2011 Study of possible systematics in the $L*_X–T*_a$ correlation of gamma-ray bursts *Astrophys. J.* **730** 135

Dainotti M G, Cardone V F, Piedipalumbo E and Capozziello S 2013a Slope evolution of GRB correlations and cosmology *Mon. Not. R. Astron. Soc.* **436** 82–8

Dainotti M G, Petrosian V, Singal J and Ostrowski M 2013b Determination of the intrinsic luminosity time correlation in the x-ray afterglows of gamma-ray bursts *Astrophys. J.* **774** 157

Dainotti M G, Del Vecchio R, Nagataki S and Capozziello S 2015a Selection effects in gamma-ray burst correlations: consequences on the ratio between gamma-ray burst and star formation rates *Astrophys. J.* **800** 31

Dainotti M, Petrosian V, Willingale R, O'Brien P, Ostrowski M and Nagataki S 2015b Luminosity–time and luminosity–luminosity correlations for GRB prompt and afterglow plateau emissions *Mon. Not. R. Astron. Soc.* **451** 3898–908

D'Avanzo P *et al* 2012 A complete sample of bright *Swift* gamma-ray bursts: x-ray afterglow luminosity and its correlation with the prompt emission *Mon. Not. R. Astron. Soc.* **425** 506–13

Efron B and Petrosian V 1992 A simple test of independence for truncated data with applications to redshift surveys *Astrophys. J.* **399** 345–52

Gehrels N *et al* 2008 Correlations of prompt and afterglow emission in *Swift* long and short gamma-ray bursts *Astrophys. J.* **689** 1161–72

Grupe D, Nousek J A, Veres P, Zhang B-B and Gehrels N 2013 Evidence for new relations between gamma-ray burst prompt and x-ray afterglow emission from 9 years of *Swift* Astrophys. *J. Suppl. Ser.* **209** 20

Kann D A *et al* 2010 The afterglows of *Swift*-era gamma-ray bursts. I. Comparing pre-*Swift* and *Swift*-era long/soft (Type II) GRB optical afterglows *Astrophys. J.* **720** 1513–58

Liang E and Zhang B 2006b Calibration of gamma-ray burst luminosity indicators *Mon. Not. R. Astron. Soc.* **369** L37–41

Lloyd N M and Petrosian V 1999 Distribution of spectral characteristics and the cosmological evolution of gamma-ray bursts *Astrophys. J.* **511** 550–61

Lloyd N M, Petrosian V and Preece R D 2000 Synchrotron emission as the source of GRB spectra, Part II: Observations *AIP Conf. Ser.* **526** 155–9

Nardini M, Ghisellini G, Ghirlanda G, Tavecchio F, Firmani C and Lazzati D 2006 Clustering of the optical-afterglow luminosities of long gamma-ray bursts *Astron. Astrophys.* **451** 821–33

Nardini M, Ghisellini G and Ghirlanda G 2008 Optical afterglows of gamma-ray bursts: a bimodal distribution? *Mon. Not. R. Astron. Soc.* **383** 1049–57

Nysewander M, Fruchter A S and Pe'er A 2009 A comparison of the afterglows of short- and long-duration gamma-ray bursts *Astrophys. J.* **701** 824–36

Oates S R *et al* 2009 A statistical study of gamma-ray burst afterglows measured by the *Swift* Ultraviolet Optical Telescope *Mon. Not. R. Astron. Soc.* **395** 490–503

Oates S R, Page M J, De Pasquale M, Schady P, Breeveld A A, Holland S T, Kuin N P M and Marshall F E 2012 A correlation between the intrinsic brightness and average decay rate of *Swift*/UVOT gamma-ray burst optical/ultraviolet light curves *Mon. Not. R. Astron. Soc.* **426** L86–90

Oates S R, Racusin J L, De Pasquale M, Page M J, Castro-Tirado A J, Gorosabel J, Smith P J, Breeveld A A and Kuin N P M 2015 Exploring the canonical behaviour of long gamma-ray bursts using an intrinsic multiwavelength afterglow correlation *Mon. Not. R. Astron. Soc.* **453** 4121–35

Panaitescu A and Vestrand W T 2008 Taxonomy of gamma-ray burst optical light curves: identification of a salient class of early afterglows *Mon. Not. R. Astron. Soc.* **387** 497–504

Petrosian V, Bouvier A and Ryde F 2009 Gamma-ray bursts as cosmological tools, arXiv: 0909.5051

Petrosian V, Singal J and Stawarz L 2013 Luminosity correlations, luminosity evolutions, and radio loudness of AGNs from multiwavelength observations *Proc. Int. Astron. Union* **9** 172

Racusin J L, Oates S R, de Pasquale M and Kocevski D 2016 A correlation between the intrinsic brightness and average decay rate of gamma-ray burst x-ray afterglow light curves *Astrophys. J.* **826** 45

Schady P, Mason K O, Page M J, de Pasquale M, Morris D C, Romano P, Roming P W A, Immler S and vanden Berk D E 2007 Dust and gas in the local environments of gamma-ray bursts *Mon. Not. R. Astron. Soc.* **377** 273–84

Yonetoku D, Murakami T, Nakamura T, Yamazaki R, Inoue A K and Ioka K 2004 Gamma-ray burst formation rate inferred from the spectral peak–energy peak luminosity relation *Astrophys. J.* **609** 935–51

Chapter 9

Redshift estimator

Given that the value of the redshift z for the majority of GRBs is not measured, recovering a correlation that is able to deduce the GRB distance from parameters independent of z would improve our knowledge of GRB features. In addition, it would provide some clues on the GRB position. For the correlations between prompt parameters, some redshift estimators were developed by Atteia (2003), Yonetoku *et al* (2004) and Tsutsui *et al* (2013). Their works consisted in numerically inverting the GRB luminosity correlations to recover the distance as a function of z. This procedure can also be implemented for the correlations between afterglow or prompt–afterglow parameters.

For example, Dainotti *et al* (2011a) developed a redshift estimator using the LT correlation. First, the best-fit slope and normalization for this correlation are determined and $\log F_{X,a}$, $\log T_{X,a}$ and $\beta_{X,a}$ are instead computed. Then, a value for z can be provided from inverting the LT correlation as is done for the correlations between prompt parameters by Yonetoku *et al* (2004). The equation (6.2) can be rewritten as:

$$
\begin{aligned}
\log L_{X,a} &= \log(4\pi F_{X,a}) + 2\log D_L(z, \Omega_M, h) - (1 - \beta_{X,a})\log(1 + z) \\
&= \log(4\pi F_{X,a}) + (1 + \beta_{X,a})\log(1 + z) \\
&\quad + 2\log r(z) + 2\log(c/H_0) \\
&= a\log\left(\frac{T_{X,a}}{1 + z}\right) + b,
\end{aligned}
\tag{9.1}
$$

where $r(z) = D_L(z, \Omega_M, h) \times (H_0/c)$. Putting on one side the quantities depending on z, it is found that

$$
\begin{aligned}
(1 + \beta_{X,a} + a)\log(1 + z) + 2\log r(z) &= a\log T_{X,a} \\
&\quad + b - \log(4\pi F_{X,a}) - 2\log(c/H_0).
\end{aligned}
\tag{9.2}
$$

For solving this equation numerically, some issues need to be considered: (1) the errors in (log $T_{X,a}$, log $F_{X,a}$, $\beta_{X,a}$) and the coefficient (a, b) of the LT correlation affect the physical quantities themselves; (2) the errors in (a, b) are not symmetric; and (3) σ_{int} is nonlinearly added to the entire error. For further details see Dainotti *et al* (2011a). The solution was applied to the E4 and the E0095 samples, indicating that the LT correlation cannot be regarded as a reliable redshift estimator. Considering $\Delta z = z_{obs} - z_{est}$, with z_{obs} and z_{est} the observed and the estimated redshifts, respectively, it was concluded that ~20% of GRBs in the E4 sample (black, $0.3 \leqslant \sigma_E \leqslant 4$, and blue, $0.095 \leqslant \sigma_E \leqslant 0.3$, GRBs in figure 9.1) recover $|\Delta z / \sigma(z_{est})| \leqslant 1$. Instead, for the E0095 set 28% of GRBs give $|\Delta z / \sigma(z_{est})| \leqslant 1$ (red GRBs in figure 9.1). The number of advantageous solutions improves in the case of $|\Delta z / \sigma(z_{est})| \leqslant 3$. Namely, it increases at ~53% (~57%) for the E4 (E0095) set, showing that σ_E has no significant effect on the evaluation of the redshift.

Large values of σ_{int} of the LT correlation is the main cause for the redshift indicator to not be reliable. Due to the large scatter of the LT correlation, it is worth testing if larger datasets will provide improved results. With this aim, they simulated an E0095 set from a distribution analogous to the E4 one. They provided (log $T_{X,a}$, $\beta_{X,a}$, z) and chose log $L_{X,a}$ from a Gaussian distribution with a mean value given by the LT correlation and variance σ_{int}. In this way, they obtained log $F_{X,a}$ and the error bars for all the quantities such that the relative errors mimicked the observations. Through Markov chains, it is demonstrated that increasing the dataset is not a proper way to improve the use of the LT correlation as a redshift estimator. Indeed, in the case $\mathcal{N} \simeq 50$, the percentage of GRBs for which $|\Delta z / \sigma(z_{est})| \leqslant 1$ first grows to ~34% and then is reduced to ~20% for $\mathcal{N} \simeq 200$. This does not come as a surprise. In fact, a large dataset provides stronger constraints on the (a, b, σ_{int}) coefficients, but not on σ_{int}. In fact, σ_{int} is the main reason for the discrepancies between z_{obs} and z_{est}.

Alternatively it was observed that, assuming the best-fit (a, b) quantities of the E0095 sample and fixing $\sigma_{int} = 0.10$ provides $f(|\Delta z / z_{obs}| \leqslant 1) \simeq 66\%$. This fact

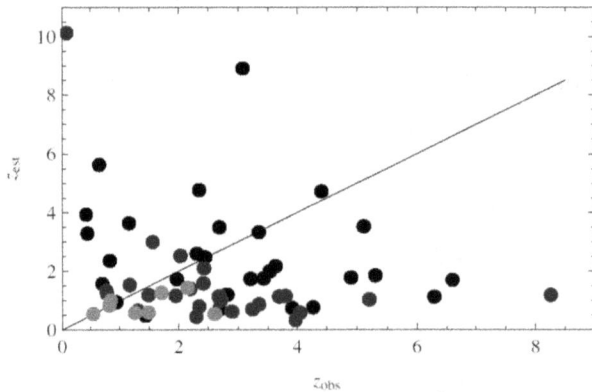

Figure 9.1. The z_{obs}–z_{est} correlation for the sample split into three σ_E intervals: $\sigma_E \leqslant 0.095$ (red dots), $0.095 \leqslant \sigma_E \leqslant 0.3$ (blue dots) and $0.3 \leqslant \sigma_E \leqslant 4$ (black dots). (Reproduced with permission from Dainotti *et al* (2011a). Copyright 2011 The American Astronomical Society.)

implied that the LT correlation can be applied as a redshift estimator only if the GRB set has $\sigma_{\text{int}} = 0.10\text{--}0.20$. Due to the small number of elements in the E0095 set, it has not been yet determined when such a GRB set can realistically be gathered. Indeed, Dainotti *et al* (2011a) found that to calibrate the LT correlation with $\sigma_{\text{int}} \sim 0.20$ and ~ 50 GRBs, ~ 600 GRBs with estimated (log $T_{X,a}$, log $F_{X,a}$, $\beta_{X,a}$, z) are required. However, it may be possible to discover some features of the afterglow phase which decrease σ_{int} of the LT correlation using a dataset that is not so large. An example of this situation is represented by the extension of the LT correlation in 3D, with the so-called fundamental plane already discussed in section 7.5. In conclusion, an interesting avenue would be to correct all the parameters of the correlations introduced in previous sections for the biases, to realize a more accurate redshift estimator.

References

Atteia J-L 2003 A simple empirical redshift indicator for gamma-ray bursts *Astron. Astrophys.* **407** L1–4

Dainotti M G, Fabrizio Cardone V, Capozziello S, Ostrowski M and Willingale R 2011a Study of possible systematics in the L^{*}_{X}–T^{*}_{a} correlation of gamma-ray bursts *Astrophys. J.* **730** 135

Tsutsui R, Yonetoku D, Nakamura T, Takahashi K and Morihara Y 2013 Possible existence of the E_p–L_p and E_p–E_{iso} correlations for short gamma-ray bursts with a factor 5–100 dimmer than those for long gamma-ray bursts *Mon. Not. R. Astron. Soc.* **431** 1398–404

Yonetoku D, Murakami T, Nakamura T, Yamazaki R, Inoue A K and Ioka K 2004 Gamma-ray burst formation rate inferred from the spectral peak energy–peak luminosity relation *Astrophys. J.* **609** 935–51

Chapter 10

Applications of GRB afterglow correlations

Here, the use of the LT correlation is presented for deriving cosmological parameters. To date, this is the only correlation between the afterglow parameters that has been employed for cosmological studies. However, the procedure is very general and can be applied to all the other correlations between the afterglow parameters. Cardone *et al* (2009, 2010) suggested the use of the afterglow phase as cosmological tool in 2009, when a new HD was obtained considering the LT correlation.

In particular, Cardone *et al* (2009) added the LT correlation to the correlations analysed in Schaefer (2007). They employed a Bayesian fitting method, analogous to the one in Firmani *et al* (2006) for the $\log E_{\gamma,\,\mathrm{peak}} - \log E_{\gamma}$ correlation. Then, they calibrated all the GRB correlations with a fiducial ΛCDM model in agreement with the data taken from the Wilkinson Microwave Anisotropy Probe, WMAP5. Analysing six correlations ($\log E_{\gamma} - \log E_{\gamma,\mathrm{peak}}$, $\log L_{\gamma,\mathrm{iso}} - \log V$, $\log L_{X,a} - \log T_{X,a}^{*}$, $\log L_{\gamma,\mathrm{iso}} - \log \tau_{\mathrm{lag}}$, $\log L_{\gamma,\mathrm{iso}} - \log \tau_{\mathrm{RT}}$ and $\log L_{\gamma,\mathrm{iso}} - \log E_{\gamma,\mathrm{peak}}$) they built a new HD including 83 GRBs (69 from Schaefer (2007) and 14 new GRBs from the LT correlation). To compute $\mu(z)$ from a dataset of 307 SNe Ia with $0.015 \leqslant z \leqslant 1.55$, a local regression was applied with the purpose of avoiding the circularity problem (see section 5.2.1). They calibrated the GRB correlations with a GRB set in the redshift range $z \leqslant 1.4$, the same interval as the SN Ia sample. This procedure allows for the use of the SN Ia set as input for the local regression of $\mu(z)$. To compute the best parameters from the regression method shown in Cardone *et al* (2009), first they simulated several datasets. The model parameters (Ω_M, w_0, w_a, h), with w_0 and w_a the coefficients of the DE EoS $w(z) = w_0 + w_a z(1 + z)^{-1}$ (Schaefer 2007), were chosen in the ranges $0.15 \leqslant \Omega_M \leqslant 0.45$, $-1.5 \leqslant w_0 \leqslant -0.5$, $-2.0 \leqslant w_a \leqslant 2.0$ and $0.60 \leqslant h \leqslant 0.80$. Then, they created a mock sample taking $\mu(z)$ from a normal distribution centred at the expected value with $\sigma_{\mathrm{int}} = 0.15$, comparable to that of the SN Ia absolute magnitude. This mock sample, with equal values of z and error distribution of the SN Ia set, is compared to the input ones. This regression procedure found reliable values of $\mu(z)$ for every cosmological model used at a given z. Afterwards, they examined the obtained HD and that from Schaefer (2007). The Schaefer HD was investigated as follows: (1) updating the parameters of the ΛCDM model, (2) applying a Bayesian fitting method and (3) also summing the LT correlation to the other correlations. To check the results of this analysis, the set of 69 GRBs used by Schaefer (2007) was also employed and, without taking into account the LT correlation, the distance moduli were calculated through the new calibration. As a result, the ratio of the new to old $\mu(z)$ is around 1 within 5%. They concluded that the new procedure did not change the outcomes in a significant way. Finally, they found that the values of $\mu(z)$ for each of the GRBs in both the Schaefer (2007) and Dainotti *et al* (2008) datasets are in agreement with those calculated in Schaefer (2007). For this reason, adding the LT correlation does not imply further systematic effects. In addition, they found that employing the LT correlation also broadens the sample from 69 to 83 GRBs and reduces the errors in $\mu(z)$ by ~14%.

Later, Cardone *et al* (2010) employed the LT correlation alone or together with additional cosmological tools to put some limits on the cosmological parameters at large z. They separated the sample of 66 LGRBs into E0095 and E4 datasets. The results indicated that the LT correlation constrains the cosmological parameters in agreement with earlier results. Indeed, the redshift domain of this current HD was broader: (0.033, 8.2). As a further step, applying the ΛCDM, the CPL (Chevallier and Polarski 2001) and the quintessence (QCDM) models, they concluded that the ΛCDM model is favoured. Then, the fit was reiterated with only SNe Ia and BAO, without GRBs, to investigate the influence of GRBs in the analysis. It was found that GRBs do not improve the estimation of the parameters, but they set the boundaries on w_a to 0. Therefore, a large set of E0095 GRBs could produce a constant DE EoS model. Unlike previous works, the HD for the E4 set is the only one obtained through a correlation between afterglow parameters with a statistically relevant subsample. Given the results obtained with the LT correlation alone, Cardone *et al* (2010) suggested that it is not necessary to sum other correlations to the LT correlation to increase the GRB sample size with computed $\mu(z)$. This is important because biases and σ_{int} are present in each correlation, and employing all the correlations together in the HD can influence the estimation of the cosmological parameters. In conclusion, Cardone *et al* (2010) pointed out that employing E0095 GRBs only, σ_{int} of the LT correlation will be severely reduced. Using the whole set of 66 LGRBs, values of Ω_M and H_0 in agreement with those available in the literature are found. Therefore, using the LT correlation for the GRB HD does not add any bias in the investigation of the cosmological quantities. Similar outcomes were obtained using E0095 GRBs only, even if they are only 12% of the dataset. Additional analysis of the E0095 sample may lead to regarding them as a standard set for investigating DE.

Following on from the above, Dainotti *et al* (2013b) investigated how much the cosmological studies are affected by a variation of 5σ (upwards and downwards) of the LT correlation slope, b, from its intrinsic value $b_{int} = -1.07^{+0.09}_{-0.14}$. They simulated through a Monte Carlo procedure a set of 101 GRBs with $b = -1.52$, $\sigma_{int} = 0.93$ (a wider value than the $\sigma_{int} = 0.66$ from the initial sample), and a ΛCDM model with

$\Omega_M = 0.291$ and $H_0 = 71$ km s^{-1} Mpc^{-1}. Then, they examined the behaviour of the scatter in the cosmological parameters for the total sample (hereafter 'Full') and a subgroup of bright GRBs with $\log L_{X,a} \geqslant 48.7$ (hereafter 'high luminosity'). This cut at a precise value of the luminosity was provided by Dainotti *et al* (2013a), who claimed that the observed luminosity for $\log L_{X,a} \geqslant 48$ is in agreement with the local luminosity function. Resembling the same approach employed by Amati *et al* (2008) for the $\log E_{\gamma,\mathrm{peak}} - \log E_{\gamma,\mathrm{iso}}$ correlation, they fit the correlation, changing both the calibration parameters, $p_{\mathrm{GRB}} = (a, b, \sigma_{\mathrm{int}})$, and the cosmological parameters, $p_c = (\Omega_M, \Omega_\Lambda, w_0, w_a, h)$, for each model. In addition to GRBs, two samples of other cosmological probes were inserted in the dataset to obtain tighter boundaries on the cosmological parameters. The first was the $H(z)$ sample $(H(z) = H_0 \times \sqrt{\Omega_M(1 + z)^3 + \Omega_k(1 + z)^2 + \Omega_\Lambda})$ in the interval $0.10 \leqslant z \leqslant 1.75$ (Stern *et al* 2010) and the second was the Union 2.1 SN Ia set composed of 580 objects in the interval $0.015 \leqslant z \leqslant 1.414$ (Suzuki *et al* 2012). Then, applying a Markov chain Monte Carlo (MCMC) and the Gelman–Rubin test[1], they tested the convergence for a cosmological model defined by particular parameters p_c to be constrained. For the full GRB sample b, a and σ_{int} of the LT correlation were independent of the cosmological model and the addition to the GRB set of the SN Ia and $H(z)$ data. Although a 5σ scatter in b_{int} is considered, these outcomes for the full sample were compatible with previous results (Dainotti *et al* 2008, 2011a), where flat models only were analysed. However, the simulated data were affected by wide errors, thus hiding the cosmological variables in the calibration. The cosmological quantities would show up only using a wider sample with low uncertainties in $(\log T_{X,a}^*, \log L_{X,a})$. In addition, to complete the study of the Full sample, a model described by the present day values of Ω_M, Ω_Λ and H_0 was used. It was employed to investigate the consequences of the deviation of b from b_{int} on the estimation of the cosmological quantities.

Even if H_0 was compatible with the results from both the local distance estimators (Riess *et al* 2009) and CMBR data (Komatsu *et al* 2011), the median Ω_M and Ω_Λ were wider than those obtained by Davis *et al* (2007) ($\Omega_M \sim 0.27$). Therefore, analysing the full sample, a different b_{int} implied a discrepancy of 13% from the best Ω_M. Although the fit of the set with SN Ia and $H(z)$ data did not indicate flat models, a flat universe was comparable with the WMAP7 cosmological parameters within 95%, obtaining $\Omega_k = -0.080^{+0.071}_{-0.093}$. However, in this case flat and not flat models were not distinguishable, even if SN data were also taken into account. For this reason, a flat model, but with the DE EoS modelled by $w(z)$, led to distinct (w_0, w_a), regardless of the presence of SN Ia and $H(z)$ samples in the analysed dataset. Instead, for the high-luminosity sample, the evaluation of the calibration parameters was not influenced by the cosmological model or whether SN Ia and $H(z)$ data are added to the set. Indeed, for the high-luminosity sample the addition of SN Ia and $H(z)$ data did not lead to an improvement in the estimation of the calibration

---

[1] This test is based on parallel chains of simulations to validate their convergence to equal posterior distribution.

parameters. In conclusion, the results for the full sample were compatible with those of the flat cosmology for SNe Ia. In constrast, for the high-luminosity sample the estimation of H_0 differed by 5% from the one computed by Petersen *et al* (2010), and the scatter in Ω_M is underestimated by 13%. They concluded that the best method for this analysis would be to employ a high-luminosity sample with a cut at $\log L_{X,a} = 48$ to avoid the addition of the luminosity and time evolutions to the cosmological parameter estimation.

In addition, Postnikov *et al* (2014) examined the DE EoS as a function of z without postulating any *a priori* $w(z)$. In this analysis, 580 SNe Ia taken from the Union 2.1 compendium (Suzuki *et al* 2012) were employed together with 54 LGRBs in the overlapping redshift area ($z \leqslant 1.4$, see the left panel of figure 10.1) to obtain a GRB HD. A standard $w = -1$ cosmological model was also considered.

In this work, the correlation coefficients of the LT correlation are derived in the range of redshift overlapping with that of SNe Ia. In this range the LT correlation reads as follows:

$$\log L_{X,a} = 53.27^{+0.54}_{-0.48} - 1.51^{+0.26}_{-0.27} \times \log T^*_{X,a}, \tag{10.1}$$

with $\rho = -0.74$ and $P = 10^{-18}$. Due to the use of the LT correlation, the redshift range was extended one order of magnitude up to $z = 8.2$.

Postnikov *et al* (2014) applied a Bayesian statistical analysis, as in Firmani *et al* (2006) and Cardone *et al* (2010). In this analysis the hypothesis is connected with a $w(z)$ function depending on H_0 and the present DE density parameter, $\Omega_{\Lambda 0}$. They assumed isotropy for the cosmological model, secure limits on the EoS and a firm value for $w(z)$ in the $z \leqslant 0.01$ range. Furthermore, they employed many random $w(z)$

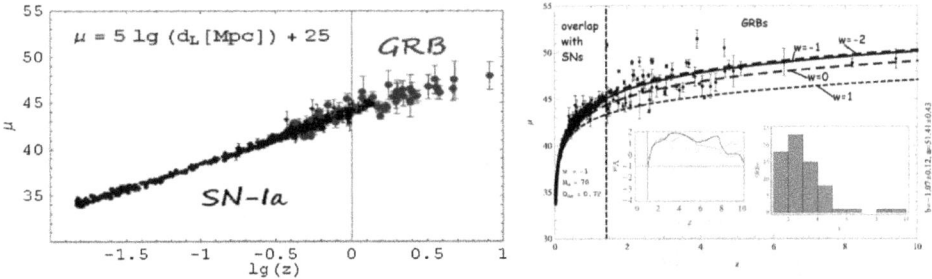

Figure 10.1. Left panel: HD for SNe Ia taken from the Union 2.1 compendium (Suzuki *et al* 2012). GRBs are obtained through the LT correlation and considering a flat $w = -1$ cosmology. Due to the significant errors in GRB data, the HD appears with no gap and compatible with the $w = -1$ model (Reproduced with permission from Postnikov *et al* (2014). Copyright 2014 The American Astronomical Society.) Right panel: HD for SNe Ia and GRBs together. GRBs in the SN Ia overlapping region are employed to compute the intrinsic correlation coefficient. Furthermore, this correlation is applied to compute $D_L(z, \Omega_M, h)$ for high-z GRBs. For comparison, constant w solutions are presented. The farthest SN Ia is indicated by a vertical dashed line. The inset panel on the left side presents the most plausible EoS with a few tested models. Due to the extended confidence intervals, from this analysis only substantial modifications from $w = -1$ can be ruled out. The inset panel on the right side displays the GRB z distribution. (Reproduced with permission from Postnikov *et al* (2014). Copyright 2014 The American Astronomical Society.)

models. To check their analysis, they simulated samples obtained through distinct input cosmological models with errors and z values equivalent to the real datasets. Employing the LT correlation, a set of *Swift* GRBs with z from 0.033 to 9.4 (see the inset in the right panel of figure 10.1) was employed to examine the history of the Universe up to $z \approx 10$. Further analysis would be helpful if the GRB at photometric redshift $z = 9.4$ were to be included in the dataset. In fact, in the sample of canonical GRBs considered by Cardone *et al* (2010) this GRB is not present. They simulated 2000 constant EoSs between $-4 \leqslant w_\Lambda \leqslant 2$. For the SN Ia sample, an exact solution was obtained. It was comparable to the cosmological constant, $w = -0.99 \pm 0.2$ (see the right panel of figure 10.1). Although the BAO boundaries provided a solution similar to that of the EoS for the SN Ia sample, it was found that the confidence interval is reduced dramatically ($w = -0.99 \pm 0.06$). Indeed, the confidence region of the solutions is relevantly constrained by BAO data. This is particularly true for $\Omega_{\Lambda 0}$, which is estimated to be $\Omega_{\Lambda 0} = 0.723 \pm 0.025$. Then, $w(z)$, which provides the best evaluation of $D_L(z, \Omega_M, h)$, z of the SN Ia set and BAO constraints, has to be chosen.

In addition to the extension of one order of magnitude in the redshift domain, GRB data should also reduce the degeneracy of the values of the cosmological parameters. Bounding the high-z $w(z)$ EoS was not straightforward, due to the small amount of data and the errors in these variables. In the left panel of figure 10.2 a simulated GRB dataset is displayed with z distribution and errors analogous to those for real data, but with an assumption of $w = -1$. It has been shown that strong $w(z)$ fluctuations alone are not possible. Then, more interesting high-z DE limits were recovered, reducing the uncertainties by 4 (see the right panel of figure 10.2). From the few GRB data, 54, in the SN Ia overlapping region it was suggested that wide errors in the GRB correlation coefficients are present. However, a shallow probability distribution (indicated by the dark area in the left panel of figure 10.2) was obtained from the wide errors in high-z GRBs for

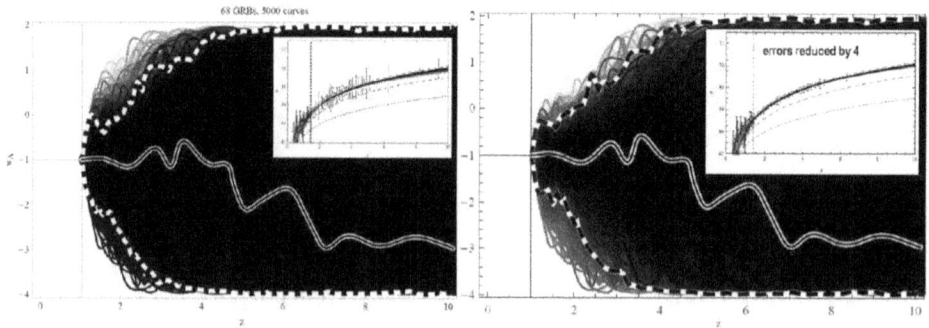

Figure 10.2. Left panel: a number of $w(z > 1)$ solutions computed through synthetic GRB datasets obtained for $w(z) = -1$. These solutions indicate how much the errors in GRBs bound the $z > 1$ EoS. The considered GRB errors are from real data. Right panel: the same as the left panel, but the assumed GRB errors are reduced by a factor of 4. (Reproduced with permission from Postnikov *et al* 2014. Copyright 2014 The American Astronomical Society.)

many analysed EoSs. For this reason, the $1 < z < 4$ region of the GRB HD will be very intriguing as soon as the GRB dataset is broad enough and the quality of data is improved.

10.1 Summary and conclusion

In this work we have reviewed GRB phenomenology, the plausible theoretical models for the mechanism behind this emission, the bivariate correlations between a number of GRB prompt and afterglow parameters and their characteristics. The scope of this work is to discuss the intrinsic nature of the thus far discovered correlations. It is important to mention that several of these correlations in the prompt emission have the problem of double truncation, which affects the parameters. In contrast, the relations involving only afterglow parameters have the advantage of not presenting double truncation in the flux limit, thus making the correction for selection effects easier and eventually allowing their use as redshift estimators and cosmological tools. Some relations have also been tested using different methodologies to prove their intrinsic nature. For others, we are not aware of their intrinsic forms and consequently how far the use of the observed relations can influence the evaluation of the theoretical models and the most relevant cosmological parameters (Dainotti *et al* 2013a), such as H_0, Ω_M, w and Ω_Λ. Therefore, the evaluation of intrinsic correlations is crucial for the determination of the most plausible model to explain the prompt and afterglow emission. In fact, although there are several theoretical explanations for each correlation, in several circumstances and under given conditions more than one interpretation is possible, thus showing that the emission processes through which GRBs are produced still need to be fully understood. To this end, it is necessary to use the intrinsic relations, and not the observed ones affected by selection biases, to test the theoretical models. These correlations might also serve as discriminating factors among different GRB classes, as several of them hold different forms for SGRBs and LGRBs, hence providing insight into the generating mechanisms. The final goal of the present effort is to present a comprehensive compendium of these correlations. This will help readers and scientists become more oriented in the vast zoo of correlations. The final aim is to push the understanding of these relations further and to use them as new standard candles eventually allowing us to explore the high-redshift universe. Indeed, a very challenging future step would be to use the corrected relations as a reliable redshift estimator and to determine further estimates of H_0, Ω_Λ and w. In other words, it is possible to employ all the relations together, with both those involving the prompt and the afterglow emission which are not yet employed for cosmological studies, as new probes, after they are corrected for selection biases, with the scope to reduce the intrinsic scatter similarly to the approach adopted by Schaefer (2007) for the prompt relations.

References

Amati L, Guidorzi C, Frontera F, Della Valle M, Finelli F, Landi R and Montanari E 2008 Measuring the cosmological parameters with the $E_{p,i}$–E_{iso} correlation of gamma-ray bursts *Mon. Not. R. Astron. Soc.* **391** 577–84

Cardone V F, Capozziello S and Dainotti M G 2009 An updated gamma-ray bursts Hubble diagram *Mon. Not. R. Astron. Soc.* **400** 775–90

Cardone V F, Dainotti M G, Capozziello S and Willingale R 2010 Constraining cosmological parameters by gamma-ray burst x-ray afterglow light curves *Mon. Not. R. Astron. Soc.* **408** 1181–6

Chevallier M and Polarski D 2001 Accelerating universes with scaling dark matter *Int. J. Mod. Phys.* D **10** 213–23

Dainotti M G, Cardone V F and Capozziello S 2008 A time–luminosity correlation for γ-ray bursts in the x-rays *Mon. Not. R. Astron. Soc.* **391** L79–83

Dainotti M G, Fabrizio Cardone V, Capozziello S, Ostrowski M and Willingale R 2011a Study of possible systematics in the $L^{*}{}_{X}$–$T^{*}{}_{a}$ correlation of gamma-ray bursts *Astrophys. J.* **730** 135

Dainotti M G, Cardone V F, Piedipalumbo E and Capozziello S 2013a Slope evolution of GRB correlations and cosmology *Mon. Not. R. Astron. Soc.* **436** 82–8

Dainotti M G, Petrosian V, Singal J and Ostrowski M 2013b Determination of the intrinsic luminosity time correlation in the x-ray afterglows of gamma-ray bursts *Astrophys. J.* **774** 157

Davis T M *et al* 2007 Scrutinizing exotic cosmological models using ESSENCE supernova data combined with other cosmological probes *Astrophys. J.* **666** 716–25

Firmani C, Ghisellini G, Avila-Reese V and Ghirlanda G 2006 Discovery of a tight correlation among the prompt emission properties of long gamma-ray bursts *Mon. Not. R. Astron. Soc.* **370** 185–97

Komatsu E *et al* 2011 Seven-year Wilkinson microwave anisotropy probe (WMAP) observations: cosmological interpretation *Astrophys. J. Suppl. Ser.* **192** 18

Petersen J H, Holst K K and Budtz-Jørgensen E 2010 Correcting a statistical artifact in the estimation of the Hubble constant based on Type Ia supernovae results in a change in estimate of 1.2% *Astrophys. J.* **723** 966–8

Postnikov S, Dainotti M G, Hernandez X and Capozziello S 2014 Nonparametric study of the evolution of the cosmological equation of state with SNe Ia, BAO, and high-redshift GRBs *Astrophys. J.* **783** 126

Riess A G *et al* 2009 A redetermination of the Hubble constant with the *Hubble Space Telescope* from a differential distance ladder *Astrophys. J.* **699** 539–63

Schaefer B E 2007 The Hubble diagram to redshift >6 from 69 gamma-ray bursts *Astrophys. J.* **660** 16–46

Stern D, Jimenez R, Verde L, Stanford S A and Kamionkowski M 2010 Cosmic chronometers: constraining the equation of state of dark energy. II. A spectroscopic catalog of red galaxies in galaxy clusters *Astrophys. J.* **188** 280–9

Suzuki N *et al* 2012 The *Hubble Space Telescope* cluster supernova survey. V. Improving the dark-energy constraints above $z > 1$ and building an early-type-hosted supernova sample *Astrophys. J.* **746** 85

www.ingramcontent.com/pod-product-compliance
Lightning Source LLC
Chambersburg PA
CBHW080547220326
41599CB00032B/6393